The FCC
AND THE
Politics of Cable TV Regulation
1952-1980

The FCC
AND THE
Politics of Cable TV Regulation

1952 - 1980

Organizational Learning and Policy Development

Michael Zarkin

Politics, Institutions, and Public Policy in America series
Series editors: Scott A. Frisch and Sean Q. Kelly

CAMBRIA PRESS

Amherst, New York

Requests for permission should be directed to:
permissions@cambriapress.com, or mailed to:
Cambria Press
20 Northpointe Parkway, Suite 188
Amherst, NY 14228

Library of Congress Cataloging-in-Publication Data

Zarkin, Michael J.
The FCC and the politics of cable TV regulation, 1952–1980 : organizational learning and policy development / by Michael J. Zarkin.
p. cm.
Includes bibliographical references and index.
ISBN 978-1-60497-722-6 (alk. paper)
1. Cable television—United States—History. 2. Telecommunication policy—United States—History. 3. United States. Federal Communications Commission—History. I. Title.

HE8700.72.U6Z37 2010
384.55'5097309045—dc22

2010030915

In loving memory of Eugene J. Frankel (1920–1992)

Table of Contents

Preface

The origins of this project can be traced back to work I began as a doctoral student in the winter of 1996 at the University of Florida. At that time, I was taking a seminar on the US Congress and desperately searching for a research design topic. While watching the news one evening, I learned that President Clinton had just signed the Telecommunications Act of 1996, a piece of legislation that was touted as a bold and far-reaching revision of US communications laws. To me, this brief news story raised important scholarly questions. How did Congress, a deliberative institution prone to policy stalemate rather than bold action, manage to enact such a sweeping change in the law? What political, social, institutional, or other types of changes had enabled Congress to pass *this* sweeping change after nearly twenty years of concerted effort? Over time, these basic queries evolved into more precise ones as I struggled to develop a theory-driven explanation for policy change. With the assistance of Larry Dodd, my doctoral advisor, I found the explanation I was looking for in a relatively new body of political science literature broadly referred to as learning theory ("social learning," "political learning," "policy learning," or "organizational learning," depending on whom you consult). The next fifteen years became a scholarly quest to apply learning theory to the study of US telecommunications policy through my dissertation, a subsequent monograph, and several research articles. The present study represents a final effort to complete this quest by developing a more precise and systematic version of the theoretical ideas I began working on nearly fifteen years ago.

This study also grew out of a desire to do something that few scholars of regulatory politics have done: conduct a detailed, in-depth, explanatory study of policy decision making by a regulatory agency. The notion that such a study is a scholarly rarity might seem controversial at first, but a close examination of the literature reveals otherwise. Nearly three generations of political scientists and economists have developed "theories of regulation" in which agencies are characterized as politicized, ineffective, inefficient, or some combination of the three. In terms of analytical depth, these studies have ranged from scholarly commentaries to highly formalized theories, but few have mustered substantial empirical evidence to back up their claims. Moreover, few of these studies have shown an appreciation for the difficulties regulatory agencies face in fulfilling their statutory missions. The policy formulation and implementation tasks undertaken by the "alphabet soup" of U.S. federal regulatory agencies—FCC, SEC, EPA, OSHA, USDA, and FDA, among others—are perhaps the most technically complex work undertaken by the civilian bureaucracy. Undoubtedly, perceptions of ineffectiveness and inefficiency occur because even the most competent agency will struggle to resolve technically complex, ill-defined, and rapidly changing policy problems. Indeed, such problems are virtually impossible to resolve with a single decision and agencies will often fail before they succeed.

The deficiencies in the regulatory politics literature discussed previously have occurred for at least two main reasons. First, social science scholarship over the last three decades has favored generalization over detailed case analysis. Thus, scholars opt for broad theoretical statements and "large-*n*" quantitative empirical studies that often miss the nuances of regulatory agency decision making. Second, studies of regulatory agency politics have tended to emphasize the role of external political forces in shaping policymaking. Although external politics is undoubtedly important, there is a second factor that scholars too often miss: internal agency decision-making processes. When scholars look inside an agency and carefully examine the process it goes through to make a major regulation, they see how personnel, technical knowledge, and ideas interact to shape policy. Admittedly, empirical exploration of internal agency processes is difficult, requiring qualitative methods (interviews, document analysis, and archival work) that many social scientists are not trained to utilize. However, by failing to account for internal agency decision-making processes, many theories of regulation do not consider the full range of variables influencing regulatory policy. In this study, I have taken advantage of the rich data set associated with the case of FCC cable television regulation in an attempt to develop a more complete picture of regulatory agency policymaking.

Whether I have ultimately succeeded or failed in accomplishing either of the goals stated above, I take full responsibility for the arguments and findings contained in this study.

Acknowledgments

Throughout the development of this project, I benefitted from the support and assistance of numerous organizations and individuals. In particular, I wish to thank Westminster College, my professional home for the past seven years, for its ongoing support. Over the past five years, this support has taken the form of a summer research grant, a semester of sabbatical, and several conference travel grants. Drs. Chuck Tripp and Gretchen Siegler, two of my Westminster faculty colleagues, deserve special mention for taking over my advising and administrative responsibilities while I was on sabbatical in the spring of 2009. The Quinney Law Library at the University of Utah, with its helpful staff and robust government documents collection, helped me get this project off the ground. The archivists at the National Archives facility in College Park, Maryland, were patient and generous with their time, providing much-needed assistance to a first-time archival researcher. The same can be said of the staff in the Lee Library at Brigham Young University, who helped me navigate the papers of FCC Commissioner Rosel Hyde. The editorial staff at Cambria Press was knowledgeable and responsive, providing timely assistance and feedback at every stage in the publication process. Comments provided by the anonymous reviewers proved invaluable during the revision process, ultimately elevating the quality of the final product. Finally, this project would never have been completed without the love, support, and encouragement of my wife, Kimberly Zarkin.

Abbreviations

ABC	American Broadcasting Company
AMST	Association of Maximum Service Telecasters
APA	Administrative Procedures Act of 1946
AT&T	American Telephone and Telegraph Company
CATV	Community Antenna Television
FCC	Federal Communications Commission
FRC	Federal Radio Commission
ICC	Interstate Commerce Commission
MIT	Massachusetts Institute of Technology
NAB	National Association of Broadcasters
NCTA	National Cable Television Association (first organized as National Community Television Association)
OTP	White House Office of Telecommunications Policy

RCA	Radio Corporation of America
UHF	Ultra High Frequency
VHF	Very High Frequency

The FCC
AND THE
Politics of Cable TV Regulation
1952-1980

Chapter 1

Introduction

The years 1952–1980 constituted a formative period in the history of US cable television regulation. During this time, the Federal Communications Commission (FCC) formulated enduring components of contemporary cable regulation including the must-carry rules, programming exclusivity rules, and the requirement that cable systems make public access channels available to local communities. Developing a workable regulatory regime for cable television, however, proved challenging. For the first decade, the FCC declined to regulate the industry, instead allowing it to mature in an unfettered marketplace. Then, in the mid-1960s, the FCC changed course, imposing regulation amid concerns that cable might cause economic harm to the broadcasting industry. In late 1968, the FCC concluded that these initial regulations were inadequate and began the process of finding a new regulatory approach. In 1972, after more than three years of proceedings, the

FCC finally implemented a comprehensive system of regulations for a cable industry that was, at this point, rapidly expanding and changing. The new regulations were a painstakingly negotiated compromise designed to promote the development of cable television while protecting broadcasters. Nevertheless, within just a few years the FCC concluded that most of these regulations were unnecessary and proceeded to repeal them.

The purpose of this study is to explain why the FCC struggled to decide whether and how to regulate the cable television industry between 1952 and 1980. From a scholarly perspective, explaining the plight of FCC cable television regulation is desirable for two reasons. First, this critical period in the history of cable regulation has never been fully analyzed from a social science perspective. Prior studies have taken the form of policy histories,[1] scholarly commentaries,[2] and theory-driven studies focused around shorter time frames.[3] None of these studies, however, employ empirical social science theory to systematically explain policy development throughout the entire twenty-eight-year period. A major purpose of the present study is to fill this gap in the literature.

Second, it is hoped that the systematic examination of FCC cable television regulation will inject new insights into the decades-old theoretical debate over the politics of regulatory agency policymaking. During the 1950s, '60s, and '70s, scholars generally adhered to the belief that regulatory agencies were susceptible to "capture" by powerful organized interests. Although capture theory produced some influential theoretical propositions, the empirical evidence was (at best) mixed.[4] Case study evidence produced in the 1980s and '90s demonstrated that US regulatory agencies may appear captured at some moments in time, whereas they act with tremendous autonomy at others.[5] Additional empirical evidence revealed that regulatory agency decision making was potentially influenced by a broad range of factors that included external political pressure, new policy ideas, technical knowledge, and the policy orientations of agency leaders.[6] Recognizing that regulatory policymaking was more complicated than once believed, political scientists and economists developed a renewed interest in formulating theories of regulation.

By the mid-1990s, two opposing theoretical camps had staked out positions as major players in the regulatory politics debate. One camp, broadly known as principal-agent theorists, developed a highly formalistic approach rooted in economic theory. According to these scholars, regulatory agency decision making is heavily influenced by the strategic choices of external political actors. The second camp, here referred to as post-positivist discourse analysts, rejected formalism in favor of a more interpretive approach focused on ideas and language. For these theorists, regulatory politics is a historically situated struggle to frame policy problems and political interests through discursive construction. Overall, both theoretical camps have contributed important insights to the understanding of contemporary regulatory politics. However, it is argued herein that neither approach has succeeded in developing an explanatory theory capable of accounting for the broad range of variables influencing regulatory policymaking.

This study does not purport to fill in all the theoretical gaps left by previous scholarship. It does, however, use the case of FCC cable television regulation to explore the explanatory power of an alternative theoretical perspective rooted in policy learning theory. The remainder of this chapter is devoted to demonstrating the relative strengths of policy learning theory as an explanatory framework. The following two sections provide an overview and critique of principal-agent and post-positivist discourse theories respectively. The final two sections present a brief overview of policy learning theory, setting the stage for further elaboration of the framework in the next chapter.

PRINCIPAL-AGENT THEORY AND REGULATION

Principal-agent theorists posit that the implementation of regulatory policy is directly influenced by political actors seeking to exert control over bureaucratic agencies. Through legislative enactments, political "principals"—legislators, interest groups, and chief executives—delegate substantial decision-making authority to bureaucratic "agents," allowing them to specify the details of regulatory policy. By delegating authority, however, principals lose direct control over policy outputs. This loss of control would not pose a problem if bureaucrats and principals indefinitely held the same interests, but they rarely do. Thus, principals must be concerned about bureaucrats taking advantage of delegated authority in order to design politically undesirable regulatory policies. Recognizing that some delegation is inevitable, principals seek out mechanisms to control or limit the choices bureaucrats make.[7]

According to principal-agent theorists, political control may be exerted through either *ex-ante* or *ex-post* mechanisms. Ex-ante control mechanisms relate to the institutional design features of the regulatory process. Principals seek to design an agency's statutory authority, organizational structure, and administrative procedures so as to ensure that bureaucratic decisions reflect established political preferences. For principals, however, the enactment of institutional design preferences is complicated by both practical and political considerations. For instance, although legislators might prefer to limit an agency's statutory authority, practical considerations—such as technical complexity—may make it necessary to delegate broad rulemaking powers.

Furthermore, principals may hold conflicting institutional design preferences, necessitating compromise among members of the enacting coalition. Thus, legislators and favored interest groups might prefer an organizational design that partially insulates the agency from the short-term agendas of chief executives (i.e., independent commission structure), but the need to engage in political compromise may force them to accept an organizational design that does not serve their long-term interests in maintaining political control (i.e., traditional executive agency structure).[8]

Nevertheless, even if legislators and interest groups fail to favorably structure statutory authority and organizational designs, they may still be able to exert ex-ante political control through administrative procedures. In a highly influential article, McCubbins, Noll, and Weingast argued that administrative procedures facilitate political control by mitigating informational disadvantages between principals and agents and by enfranchising important constituencies. General procedural frameworks, like the Administrative Procedures Act (APA) of 1946, reduce information disparities by forcing agencies to undertake decision making in a very public atmosphere. Principals may also stack the deck in favor of powerful groups by including burden-of-proof standards, judicial review limitations, and other types of specialized procedures in an agency's enabling legislation.[9]

Administrative procedures also carry the benefit of providing principals with the information necessary to successfully employ ex-post control strategies—techniques principals use to monitor and influence agency actions long after design decisions have been made. For instance, legislative oversight is one major type of ex-post control that has been studied extensively by principal-agent scholars. McCubbins and Schwartz contended that legislative oversight exists in two varieties: "police patrols" and "fire alarms." When legislators engage in police-patrol oversight, they act on their own initiative to continuously monitor the work of agencies. When legislators engage in fire-alarm oversight, they respond on a more ad hoc basis to concerns raised by citizens and interest groups.[10] Using a cost/benefit calculus, McCubbins and Schwartz originally posited that legislators would be more likely to employ a fire-alarm approach, devoting time and resources to oversight only when it was politically advantageous. However, subsequent research demonstrated that legislators and their staff devoted substantial time and energy to monitoring bureaucratic activities on an ongoing basis.[11] Therefore, principal-agent theorists now readily accept the proposition that police-patrol oversight is more beneficial to principals than originally believed.[12]

Other scholars have asserted that ex-post controls consist of a broader repertoire of mechanisms that includes oversight, budget appropriations, and the selection of agency heads. Principals use a combination of these mechanisms to create incentives for regulatory agencies to pay heed to the principals' own political preferences. In their study of the Federal Trade Commission, Weingast and Moran concluded that throughout the 1970s, Congress employed hearings, appointments, and budgetary authority in order to exert continuous control over the scope and direction of agency policy.[13] Research by Terry Moe, as well as work by Wood and Waterman, demonstrated that presidents have also employed ex-post controls and that the appointment of agency heads is a particularly effective way for a chief executive to change the trajectory of regulatory policy.[14]

Overall, principal-agent studies have broadened scholarly understanding of regulatory agencies. In particular, these studies have provided systematic theoretical accounts of how external political actors attempt to strategically influence regulatory agency decision making. Principal-agent studies have also strongly suggested that institutional design plays a role in determining how regulatory agencies operate. These general conclusions are not challenged in this study. For the reasons that follow, however, it is argued here that principal-agent models do not provide a general theory of regulatory agency policymaking that can be used to explain the case of FCC cable television regulation.

First, although principal-agent scholars have constructed elegant theoretical models, they have provided (at best) modest empirical evidence to support the proposition that political control strategies actually influence agency decision making.[15] Furthermore, the few empirical studies that have been conducted focus almost exclusively on agency enforcement actions rather than on the making of regulations.[16] Enforcement actions tend to be routine activities that may be more easily influenced by political control strategies. By contrast, regulations are generally applicable legal provisions made by agencies that possess extensive technical knowledge within their policy domains. Thus, there is reason to hypothesize that the making of regulations will be less pervious to political control than an agency's enforcement actions. Indeed, the few studies that have closely examined the impact of principal-agent relationships on the making of regulations have found meager evidence that political control strategies are successful.[17]

Second, principal-agent theorists unrealistically assume that principals will employ control strategies in defense of a fixed and well-defined set of political preferences. This assumption is problematic because it does not allow for situations in which principals either do not have clear preferences or do not have the foresight to predict future preferences. Indeed, the case of cable television provides an illustration of how unpredictable political preferences can be. When the Communications Act of 1934 was passed, Congress had no idea that the cable television industry would someday exist. Furthermore, at that time, television broadcasting was still in its experimental stages and could not have been characterized as a fully formed industry, let alone a politically influential one. As a result, it would have been very difficult—if not impossible—for Congress to act strategically in favor of either the broadcasting or the cable industry.[18] Furthermore, once the practical and political implications of cable television had become clear in the 1960s, Congress made no serious effort to drastically alter the FCC's statutory authority or to impose procedural constraints that would advantage either side of the controversy. Rather, Congress seemed willing, if not content, to let the FCC gain experience with cable television and experiment with new policy choices.

Finally, principal-agent models do not provide a complete explanation for regulatory agency policymaking because they preclude consideration of factors other than the strategic decisions of external political actors. With respect to the FCC, at least, a more thorough explanation of policymaking requires consideration of the internal workings of the agency itself as well as of its external political and institutional environment. The FCC is an independent regulatory commission entrusted by Congress with a broad delegation of statutory authority. External actors—such as presidents, legislators, interest groups, and courts—create legal and political opportunities and constraints that frame the FCC's actions; however, these external forces do not control the FCC's decisions in any real sense. Rather, the scope and direction of FCC regulatory policy is determined equally by the agency's ability to access and utilize technical knowledge, by the policy orientations of FCC leaders, and by the broader policy ideas and beliefs that frame discussions concerning US communications policy.[19]

Thus, in order to provide a complete explanation of regulatory policymaking, a theory must take into consideration the roles played by knowledge and ideas within the policy process. Although they are not generally considered important variables in principal-agent models, policy ideas are the central focus of discourse theory. The following section provides an overview and a critical assessment of discourse perspectives as a theory of regulatory policymaking.

Ideas, Discourse, and Regulation

Political scientists have long recognized the importance of ideas within the policy process. Only recently, however, have ideas been accorded any intrinsic value as a force for policy change. Prior to the mid-1990s, most policy scholars believed that ideas, though at times rhetorically powerful, were a direct outgrowth of strategic political interests.[20] Political actors intentionally used "heresthetics,"[21] "policy images,"[22] or "social constructions"[23] to gain support for their preferred policy options. Ideas and issue frames facilitated coalition building by simplifying reality for political actors who possessed inherently limited information processing capabilities. Nevertheless, ideas were little more than symbolic shorthand for otherwise objective social conditions. As social change occurred, ideas necessarily became outmoded and political coalitions became tenuous. Social change thus provided strategic political actors with an opportunity to build new coalitions around new issue frames.

More recently, however, scholars in the post-positivist tradition have argued that ideas are not simply a reflection of interests but a constitutive force in and of themselves. Generally speaking, post-positivists reject the notion that social and political realities can be understood in objective terms; instead, they view the policy process as a highly interpretive and value-laden struggle to define problems through language and discourse.

Social problems and phenomena are given meaning through the linguistic concepts employed by political actors who view the world through differing mental frames.[24] Language thus becomes an integral component of reality, shaping the prospects of political change by serving as a medium for the definition of interests and the construction of new meanings and identities.[25] Through argumentation, opposing political actors compete to define reality by presenting metaphors and discursive narratives that serve to unite diverse intellectual orientations around a common symbolic conception of a policy problem.[26] The choice of a uniting discursive construct or idea, however, is not a purely strategic decision. Rather, institutional arrangements, historical context, and prevailing societal values frame the choices available to political actors who are seeking to discursively define policy problems.[27]

The post-positivist approach to ideas is particularly relevant to this study for two reasons. First, post-positivist theoretical approaches have been applied extensively to the study of regulatory politics and policy in recent years.[28] The second, and perhaps more important, reason is that two of the main theory-driven studies of FCC cable television regulation have employed post-positivist discourse analysis. The remainder of this section provides an overview of these two studies and a critical assessment of the strengths and weaknesses of post-positivist discourse analysis as a tool for studying regulatory policymaking.

In an influential 1987 article, Thomas Streeter argued that the FCC's 1972 cable television regulations were shaped by a loosely linked but politically powerful discourse.[29] Streeter located the underlying themes of this discourse in a series of journal articles and reports published between 1966 and 1970 by academics, think tanks, government study commissions, and organized interests. Taken together, these studies framed cable television as a revolutionary technology capable of expanding access to education, improving democratic deliberation, and generally facilitating the resolution of numerous social problems. Over time, the broad and inclusive themes associated with the cable discourse captured the imaginations of FCC officials and of a diverse array of organized interests, creating a shared language through which problem solving could be approached. According to Streeter, the new discourse impacted the content of the 1972 rules in key respects; for example, the FCC required cable systems to make public access, educational, and governmental channels available at the community level.

Within two years of the publication of Streeter's argument, Patrick Parsons published an article in which he argued that discursive framing had also influenced the FCC's cable television deliberations in the 1950s and early 1960s.[30] According to Parsons, cable television's early years were a time when cable and broadcast interests were competing to frame the new technology in ways that advantaged their own policy preferences. In these efforts, however, each side was forced to navigate the opportunities and constraints presented by norms, legal structures, and the existing balance of political power. Broadcasters sought regulation of cable television in an effort to avoid competition. Under existing legal arrangements, the most effective way to advance this position was to portray cable as a direct extension of the broadcast system; as such, cable would be immediately subject to regulation by the FCC. Members of the cable industry, by contrast, sought to avoid FCC regulation and adherence to copyright laws. With these goals in view, they chose to present cable as little more than an extension of the viewer's television antenna and thus not subject to regulation under existing laws. By the early 1960s, the broadcasters' frame had prevailed, in large part because the balance of political power shifted in their favor.

At that time, most of the FCC commissioners and many of the most powerful broadcasters were predisposed to regulation. The broadcasters' frame thus provided a powerful intellectual road map for the advancement of a preferred course of action.

By focusing scholarly attention on ideas and discourse, both Streeter and Parsons contributed important insights to the understanding of both FCC policy deliberations and regulatory policymaking in general. In particular, the two studies demonstrate that ideas, interests, and institutional arrangements can interact in complex ways to shape policy change. Although political actors may appear to intentionally select issue frames that support their strategic interests, their choices are limited by existing legal frameworks and other contextually relevant factors. At the same time, however, the choice of issue frames can have a constitutive effect, redefining issues in ways that can alter perceived interests and point the way to previously unforeseen policy options. These findings are by no means dismissed in the present study. Indeed, the learning model presented in the next chapter shares many of these basic convictions concerning the role of ideas in the policy process.

Nevertheless, the discourse analyses presented by Streeter and Parsons provide an incomplete picture of FCC policy deliberations for at least two reasons. First, both studies provide a highly limited characterization of how ideas shape policymaking. In particular, both studies focus on the narrow issue frames offered by contending interests without examining whether these discursive constructs might be related to much broader, historically situated sets of policy ideas. For instance, Streeter convincingly describes the rhetorical success of the cable discourse during the late 1960s, but he does not provide an analytical framework capable of explaining how this issue frame relates to the broader "public interest" goals that had undergirded communications policy for decades.[31] By contrast, Parsons' approach, with its emphasis on historical norms and institutional arrangements, seems inclusive enough to incorporate a broader discussion of policy ideas. Unfortunately, Parsons' empirical analysis focuses narrowly on legal frameworks and does not explore the ideas upon which those frameworks are based.

Second, the Streeter and Parsons studies both suffer from a limitation that plagues post-positivist discourse frameworks in general: a failure to appreciate the unique and privileged role that "the state" plays in the policy process. Post-positivists tend to characterize the policy process as an exchange of ideas within a discourse community consisting of politicians, civil servants, organized interests, and other privileged intellectual elites. Although the preferences of state actors, such as FCC officials, are by no means dismissed, neither are they accorded any special analytical significance. For the purposes of this study, however, a close examination of state decision-making activities is essential. Because state actors possess formal decision-making authority, their preferences and their knowledge about the world are the forces that most immediately shape the content of public policy. The knowledge held by state actors may include broad, historically situated frames, but it also includes experience accrued through practical problem solving. Practical experience becomes particularly important when state decision makers confront a novel, technically complex policy problem, such as that presented by cable television. When confronted with new policy problems, decision makers may attempt to simplify matters by framing them in light of analogies drawn from past experience. When problems are complex and multifaceted, however, framing by analogy may prove insufficient as a road map for their resolution. Consequently, problems of this kind may take years or even decades to resolve, as state decision makers experiment with solutions, accrue knowledge, and reorient their mental frames. As decision makers undertake this process, formal state organizations—through their personnel, precedents, and norms—become the storehouses for accrued practical policy knowledge.

The pattern outlined in the previous paragraph is a pattern the FCC has repeated many times while integrating various new technologies into the US communications infrastructure. In each instance—cable television, FM radio, television, computers, and broadband are just a few—the FCC's earliest regulatory actions took place in an environment characterized by limited or ambiguous information. As a result, none of these policy problems were resolved through a single decision at a single moment in time. Rather, decisions were made, assessed, and revised as new information became available. Along the way, the FCC acquired knowledge and skills it had not previously possessed. Thus, although many individual FCC decisions have been judged a failure, over time, the agency's ability to change course and seek out new policy solutions has allowed the US communications industry to flourish. Cable television is no exception to this pattern.

LEARNING AND POLICYMAKING: A CONCEPTUAL OVERVIEW

The foregoing discussion suggests that neither principal-agent theory nor post-positivist discourse analysis can adequately explain the case of early FCC cable television regulation. What is needed is a theoretical perspective that does not dismiss political power, but that explains how political actors and institutional arrangements both constrain and enable effective bureaucratic decision making. Such a theory must effectively explain how agency leadership, ideas, and technical knowledge influence decision making in a complex and uncertain policy environment. Finally, the theory needs to address policymaking as a developmental process that unfolds over time. In the remainder of this chapter, it is argued that an approach rooted in policy learning theory can meet these theoretical specifications. This section presents a broad conceptual overview of policy learning theory; in the final section of the chapter, it is argued that FCC cable television regulation can be best analyzed through a specific focus on organizational learning.

The notion that policymaking may be characterized as a learning process was first articulated by Hugh Heclo in his study of social politics in Britain and Sweden. After reviewing the historical progression of social welfare policy in these countries, Heclo concluded that policy change cannot be explained simply by shifts in the locus of power among competing political forces. Societal change, elections, and group mobilization unquestionably produce shifts in influence and bring new demands before the state. These forces, however, are best understood as environmental stimuli that spur policymakers into action. Once begun, policy formulation is as much an intellectual as a political process, and actors compete to define the nature of problems and to propose appropriate solutions. As Heclo noted in a frequently quoted passage:

> Politics finds its sources not only in power but also in uncertainty—men collectively wondering what to do. Finding feasible courses of action includes, but is more than, locating which way the vectors of political pressure are pushing. Governments not only "power" (or whatever the verb form of that approach might be); they also puzzle. Policy-making is a form of collective puzzlement on society's behalf; it entails both deciding and knowing.[32]

Heclo characterized the policy changes resulting from this process of puzzlement as political learning.

Drawing inspiration from cognitive psychology, Heclo defined learning as "a relatively enduring alteration in behavior that results from experience."[33] In essence, policymaking participants learn lessons from past successes and failures. Over time, policymakers build up a repertoire of workable policy techniques that can be matched to new situations as circumstances dictate. Intellectually speaking, policymakers prefer to stick to the familiar, employing proven techniques whenever they are easily applied to new problems. However, when new problems are ambiguous, or when policy failure demonstrates that favored techniques are unworkable, a window of opportunity opens for greater innovation. At these moments, policy "middlemen" spanning the boundary between state and society have the opportunity to advocate new policy formulations. These political actors, however, bear a heavy intellectual burden and must employ a combination of expert knowledge and argumentation in order to build support for favored initiatives.

In advancing the political learning thesis, Heclo opened a new theoretical avenue for scholars interested in exploring the intellectual dimensions of policy change. However, Heclo also left many theoretical questions unanswered, leaving future researchers to debate the scope and nature of learning within the policy process. In particular, scholarship is divided on the question of who learns—that is, which actors or institutions within the policymaking process are the vehicles for learning? Studies of US foreign policy and British macroeconomic policy suggest that state leaders and their closest advisors are the primary vehicles for learning.[34] These findings contrast with those of Sabatier and Jenkins-Smith, who argued that learning is a discursive process involving a broad range of state and societal actors within specific policy domains.[35] Still other scholars have argued that governmental bodies may engage in "organizational learning" by incorporating new experiences and knowledge into their organizational memories.[36]

To some extent, this is a misplaced controversy. In any given policy area, various actors and organizations either learn or supply knowledge to those who learn. Thus, the real question is not who learns, but rather who needs to learn in order for policy to change? Accumulated evidence from numerous empirical studies has made clear that the agents of learning vary depending on the locus of decision-making authority within a particular policy domain.[37] For instance, foreign policy and macroeconomic policy are matters of system-wide importance typically entrusted to high-ranking national executives. In issue areas such as these, it makes sense that small groups of national leaders are the principal learners. In an area such as telecommunications regulation, however, policy questions are technically complex and only intermittently salient to the mass public. As a result, elected officials delegate substantial authority to administrative agencies, making organizational learning the norm in this context. In an issue area such as environmental policy, on the other hand, major policy disputes tend to be both complex and salient. The general public frequently expresses concern regarding the impacts of environmental policy, and scientific knowledge is fiercely contested among various competing interest groups. Legislators tend to delegate less decision-making responsibility to administrative agencies in these cases, and they themselves remain actively involved in the policy process. Under these circumstances, a range of state and societal actors must learn before there can be sufficient support for policy change.[38]

In addition to debating *who* learns, learning scholars have also sought to clarify *what* is learned. Most scholars agree that learning is indicated when policy change results directly from new knowledge and experience. Thus, policy change that is attributable only to political pressure or perceived self-interest cannot be categorized as learning. It is likewise accepted among scholars that the scale of policy change is related to the scale of knowledge acquisition. For instance, new research and experience may lead to relatively incremental policy changes without altering established policy goals. By contrast, broad-based policy changes involve a reconsideration of fundamental goals and are often accompanied by a more profound shift in knowledge—a shift that alters the underlying ideas and assumptions that guide decision making.[39]

More recently, scholars have also sought to understand when learning is most likely to occur. In issue areas that are politically salient but not technically complex, policymakers tend to follow majoritarian preferences rather than rely solely on expert knowledge or practical experience. Issue areas such as these are generally not conducive to learning.[40] By contrast, learning is more likely in technically complex, rapidly changing issue areas—situations in which policymakers are required to step away from their political instincts and seek out new knowledge. Yet, even in these issue areas, learning is only possible when policymakers are open to new ideas, when they have the means to gather and analyze new knowledge, and when the institutional and political environment is conducive to learning.[41]

Cable Regulation as Organizational Learning?

This study explores the proposition that the FCC needed to learn in order to regulate cable television. For the purposes of this analysis, a theoretical approach rooted in the organizational learning literature is employed—this focus is appropriate for two broad reasons. First and foremost, the FCC was the primary decision-making agent in the case under examination. In fashioning regulations for cable television, the FCC drew on a broad delegation of authority granted under the Communications Act of 1934. During the period considered here, this grant of authority was not substantially altered by either the Congress or the US courts. Thus, if learning did contribute to policy change, then that learning necessarily took place within the organizational walls of the FCC.

Second, organizational learning theory offers a framework well suited to explaining many of the features (discussed earlier in this chapter) of FCC decision making. Organizational learning theory allows the analyst to consider the role played by experience and expert knowledge in highly technical areas of public policy. Furthermore, organizational learning theory provides an integrative perspective that can explain how institutional arrangements, external political forces, and internal agency dynamics combine to influence regulatory agency policymaking. Finally, because it emphasizes the accumulation of experience and new knowledge over time, organizational learning theory is appropriate in an examination of policy change over the course of several decades.[42]

The remainder of this study is dedicated to outlining and applying the organizational learning thesis to the case of FCC cable television regulation. Chapter 2 develops the thesis, provides further justification for its application to the case of cable television, and outlines a series of research questions to guide the analysis. In chapters 3–7, in-depth case study evidence is presented to test the propositions outlined in chapter 2. Chapter 8—the concluding chapter—contains an analysis of the research findings, a postscript that briefly outlines important developments in cable regulation after 1980, and some concluding thoughts concerning the value of organizational learning theory for the broader study of regulatory agencies.

ENDNOTES

1. Stanley M. Besen and Robert W. Crandall, "The Deregulation of Cable Television," *Law and Contemporary Problems* 44 (1981): 77–124; Patrick R. Parsons, *Blue Skies: A History of Cable Television* (Philadelphia: Temple University Press, 2008).
2. Martin H. Seiden, *Cable Television USA: An Analysis of Government Policy* (New York: Praeger, 1972); Don R. LeDuc, *Cable Television and the FCC: A Crisis in Media Control* (Philadelphia: Temple University Press, 1973).
3. Thomas Streeter, "The Cable Fable Revisited: Discourse, Policy, and the Making of Cable Television," *Critical Studies in Mass Communication* 4 (1987): 174–200; Patrick R. Parsons, "Defining Cable Television: Structuration and Public Policy," *Journal of Communication* 39, no. 2 (1989): 10–26.
4. Marver H. Bernstein, *Regulating Business by Independent Commission* (Princeton, NJ: Princeton University Press, 1955); Theodore J. Lowi, *The End of Liberalism* (New York: W. W. Norton, 1969). Here, I am referring mainly to the capture theorists who wrote about regulatory agency behavior. Economic capture theory—more commonly known as the economic theory of regulation—systematically examines the origins of regulation in general, encompassing the behavior of legislators as well as bureaucrats. On the economic theory of regulation, see George J. Stigler, "The Theory of Economic Regulation," *Bell Journal of Economics and Management Science* 2 (1971): 3–31. Economic capture theory has received some empirical support in studies of regulatory policy at the state level, though most scholars have concluded that interest groups are only one variable influencing regulatory policy. See, for instance, William Gormley, *The Politics of Public Utility Regulation* (Pittsburgh: University of Pittsburgh Press, 1983); Kenneth J. Meier, *The Political Economy of Regulation: The Case of Insurance* (Albany: State University of New York Press, 1988); Paul Teske, *After Divestiture: The Political Economy of Regulation* (Albany: State University of New York Press, 1990).

5. Gregory Hooks, "From an Autonomous to a Captured State Agency: The Decline of the New Deal in Agriculture," *American Sociological Review* 55 (1990): 29–43; Christopher McGrory Klyza, "A Window of Autonomy: State Autonomy and the Forest Service in the Early 1900s," *Polity* 25 (1992): 173–196; Erwin G. Krasnow, Lawrence D. Longley, and Herbert A. Terry, *The Politics of Broadcast Regulation*, 3rd ed. (New York: St. Martin's Press, 1982); Martha Derthick and Paul J. Quirk, *The Politics of Deregulation* (Washington, DC: Brookings Institution, 1985).
6. Richard A. Harris and Sidney M. Milkis, *The Politics of Regulatory Change: A Tale of Two Agencies* (New York: Oxford University Press, 1989); Marc K. Landy, Marc J. Roberts, and Stephen R. Thomas, *The Environmental Protection Agency: Asking the Wrong Questions* (New York: Oxford University Press, 1990); Marc Allen Eisner, *Antitrust and the Triumph of Economics: Institutions, Expertise, and Policy Change* (Chapel Hill: University of North Carolina Press, 1991); Hugh R. Slotten, *Radio and Television Regulation: Broadcast Technology in the United States, 1920–1960* (Baltimore: Johns Hopkins University Press, 2000).
7. Mathew D. McCubbins, "The Legislative Design of Regulatory Institutions," *American Journal of Political Science* 29 (1985): 721–748.
8. Terry M. Moe, "The Politics of Bureaucratic Structure," in *Can the Government Govern?*, ed. John E. Chubb and Paul E. Peterson, 267–329 (Washington, DC: Brookings Institution, 1989); David Epstein and Sharyn O'Halloran, *Delegating Powers* (Cambridge: Cambridge University Press, 1999); B. Dan Wood and John Bohte, "Political Transaction Costs and the Politics of Administrative Design," *Journal of Politics* 66 (2004): 176–202.
9. Mathew D. McCubbins, Roger G. Noll, and Barry R. Weingast, "Administrative Procedures as Instruments of Political Control," *Journal of Law, Economics, and Organization* 3 (1987): 243–277. Several studies have elaborated the McCubbins-Noll-Weingast thesis. See Kathleen Bawn, "Political Control versus Expertise: Congressional Choices about Administrative Procedures," *American Political Science Review* 89 (1995): 62–73; Charles R. Shipan, *Designing Judicial Review: Interest Groups, Congress, and Communications Policy* (Ann Arbor: University of Michigan Press, 1997).

10. Mathew D. McCubbins and Thomas Schwartz, "Congressional Oversight Overlooked: Police Patrols versus Fire Alarms," *American Journal of Political Science* 28 (1984): 165–179.
11. See, in particular, Joel D. Aberbach, *Keeping a Watchful Eye* (Washington, DC: Brookings Institution, 1990).
12. Arthur Lupia and Mathew D. McCubbins, "Learning from Oversight: Fire Alarms and Police Patrols Reconstructed," *Journal of Law, Economics, and Organization* 10 (1994): 96–125.
13. Barry R. Weingast and Mark Moran, "Bureaucratic Discretion or Congressional Control?" Regulatory Policymaking by the FTC," *Journal of Political Economy* 91 (1983): 765–800.
14. Terry M. Moe, "Control and Feedback in Economic Regulation: The Case of the NLRB," *American Political Science Review* 79 (1985): 1094–1116; B. Dan Wood and Richard W. Waterman, "The Dynamics of Political Control of the Bureaucracy," *American Political Science Review* 85 (1991): 801–828.
15. John D. Huber and Charles R. Shipan, two theorists in the ex-ante control tradition, concede as much. As they note in a recent review essay, "No one has yet been able to solve the very difficult problem of assessing empirical relationships between the preferences of bureaucrats and politicians on the one hand and policy outcomes on the other." See their "Politics, Delegation, and Bureaucracy," in *The Oxford Handbook of Political Economy*, ed. Barry R. Weingast and Donald A. Wittman, 261 (Oxford: Oxford University Press, 2006).
16. By enforcement actions, I mean routine agency activities designed to implement a broader law or regulation. Examples include the issuing of permits and licenses, tickets and fines, and the filing of lawsuits. For a discussion of this deficiency in the literature, see David B. Spence, "Agency Policymaking and Political Control: Modeling Away the Delegation Problem," *Journal of Public Administration Research and Theory* 7 (1997): 212–213.
17. A good example is the work of Ferejohn and Shipan. In their studies of Congressional influence on FCC telephone regulation, they prefer the term "influence" to "control." In the case studies they examined, Congress held numerous hearings and threatened legislative action, but ultimately succeeded in obtaining only minor alterations in regulations. See John A. Ferejohn and Charles R. Shipan,

"Congressional Influence on Administrative Agencies: A Case Study of Telecommunications Policy," in *Congress Reconsidered*, 4th ed., ed. Lawrence C. Dodd and Bruce I. Oppenheimer, 393–410 (Washington, DC: Congressional Quarterly Press, 1989); "Congress and Telecommunications Policymaking," in *New Directions in Telecommunications Policy*, ed. Paula Newburg, 301–314 (Durham, NC: Duke University Press, 1989).

18. McCubbins, Noll, and Weingast seem to think otherwise. In their seminal 1987 article, they argue that the FCC made cable television regulations on "autopilot" during the 1960s and '70s. In essence, the regulatory process was set up to automatically reflect prevailing political preferences without any interference by Congress; see McCubbins, Noll, and Weingast, "Administrative Procedures," 271. However, Glen O. Robinson, a former FCC commissioner, seriously doubts their thesis. See his "Commentary on 'Administrative Arrangements and the Political Control of Agencies': Political Issues of Structure and Process," *Virginia Law Review* 75 (1989): 491–492.
19. I tease this conclusion out of a number of studies of FCC decision making conducted by political scientists, historians, and legal scholars. See, for instance, Derthick and Quirk, *The Politics of Deregulation*; Alan Stone, *Wrong Number: The Breakup of AT&T* (New York: Basic Books, 1989); Michael J. Zarkin, *Social Learning and the History of Telecommunications Policy, 1900–1996* (Lewiston, NY: Edwin Mellen Press, 2003); Michael J. Zarkin, "Organizational Learning in Novel Policy Situations: Two Cases of United States Communications Regulation," *Policy Studies* 29 (2008): 87–100; Slotten, *Radio and Television Regulation*; Wenmouth Williams, Jr., "Impact of Commissioner Background on FCC Decisions, 1962–1975," *Journal of Broadcasting* 20 (1976): 239–260; LeDuc, *Cable Television and the FCC*.
20. For instance, John Kingdon defines ideas in part as "theories that individuals hold about how the world works … that can be used to further self-interest as well as other goals." See his "Agendas, Ideas, and Policy Change," in *New Perspectives on American Politics*, ed. Lawrence C. Dodd and Calvin Jillson, 221 (Washington, DC: Congressional Quarterly Press, 1994).

21. William H. Riker, "The Heresthetics of Constitution-Making: The Presidency of 1787, with Comments on Determinism and Rational Choice," *American Political Science Review* 78 (1984): 1–16.
22. Frank R. Baumgartner and Bryan D. Jones, *Agendas and Instability in American Politics* (Chicago: University of Chicago Press, 1993).
23. Anne L. Schneider and Helen M. Ingram, *Policy Design for Democracy* (Lawrence: University of Press of Kansas, 1997).
24. Frank Fischer, *Reframing Public Policy: Discursive Politics and Deliberative Practices* (Oxford: Oxford University Press, 2003), 41–44, 81–88.
25. Maarten A. Hajer, *The Politics of Environmental Discourse: Ecological Modernization and the Policy Process* (Oxford: Oxford University Press, 1995), 59.
26. Ibid., 62; Donald Schon, "Generative Metaphor: A Perspective on Problem-Setting in Social Policy," in *Metaphor and Thought*, 2nd ed., ed. Andrew Ortony, 137–163 (Cambridge: Cambridge University Press, 1993).
27. Thomas Streeter, *Selling the Air: Commercial Broadcasting in the United States* (Chicago: University of Chicago Press, 1996), 115; Herbert Gottweis, *Governing Molecules: The Discursive Politics of Genetic Engineering in Europe and the United States* (Cambridge, MA: MIT Press, 1998), 33–36.
28. See particularly Hajer, *Politics of Environmental Discourse*; Streeter, *Selling the Air*; Gottweis, *Governing Molecules*. See also Karen T. Litfin, *Ozone Discourses: Science and Politics in Global Environmental Cooperation* (New York: Columbia University Press, 1994); Bruce A. Williams and Albert R. Matheny, *Democracy, Dialogue, and Environmental Disputes: The Contested Languages of Social Regulation* (New Haven, CT: Yale University Press, 1995).
29. Streeter, "The Cable Fable."
30. Parsons, "Defining Cable Television."
31. More recent post-positivist writings have discussed how broad "meta-narratives" may shape policy deliberations at any given moment. Indeed, in later work, Streeter demonstrated how a broad meta-narrative known as "corporate liberalism" shaped the development of US communications policy during the early 20th century.

See Streeter, *Selling the Air*, 94–140. These meta-narratives, however, are a much broader species of ideas than those considered in this study.

32. Hugh Heclo, *Modern Social Politics in Britain and Sweden: From Relief to Income Maintenance* (New Haven, CT: Yale University Press, 1974), 305.
33. Ibid., 306.
34. Jack S. Levy, "Learning and Foreign Policy: Sweeping a Conceptual Minefield," *International Organization* 48 (1994): 279–312; David Patrick Houghton, "The Role of Analogical Reasoning in Novel Foreign Policy Situations," *British Journal of Political Science* 26 (1996): 523–552; Michael J. Oliver, *Whatever Happened to Monetarism?: Social Learning and Macroeconomic Policymaking in the United Kingdom since 1979* (Aldershot, U.K.: Ashgate, 1997).
35. Paul A. Sabatier and Hank C. Jenkins-Smith, eds., *Policy Change and Learning: An Advocacy Coalition Approach* (Boulder, CO: Westview Press, 1993).
36. M. Leann Brown, Michael Kenney, and Michael Zarkin, *Organizational Learning in the Global Context* (Aldershot, U.K.: Ashgate, 2006).
37. Ibid. This conclusion emerges from a broad reading of the studies in Brown, Kenney, and Zarkin, eds., *Organizational Learning in the Global Context.*
38. Here, I simply draw on the old adage that "policy determines politics." Many scholars have developed "policy typologies" in which they suggest that the nature of any given issue area will determine the structure of interest representation and institutional arrangements. See, in particular, Theodore J. Lowi, "American Business, Public Policy, Case Studies, and Political Theory," *World Politics* 16 (1964): 687–691; James Q. Wilson, *Political Organizations* (New York: Basic Books, 1973); William T. Gormley, "Regulatory Issue Networks in a Federal System," *Polity* 18 (1986): 595–620.
39. Peter A. Hall, "Policy Paradigms, Social Learning, and the State: The Case of Economic Policymaking in Britain," *Comparative Politics* 25 (1993): 275–296; M. Leann Brown, "Scientific Uncertainty and Learning in European Union Environmental Policymaking,"

Policy Studies Journal 28 (2000): 576–596; Michael J. Zarkin, "Telecommunications Policy Learning: The Case of the FCC's Computer Inquiries," *Telecommunications Policy* 27 (2003): 283–299.

40. Issues that fit this category include conflicts over morality and the American way of life. Most prominent among these are questions that relate to constitutional rights such as pornography, gun control, and affirmative action. Immigration, the minimum wage, and truth in advertising are similarly situated issues. See Gormley, "Regulatory Issue Networks," 600.
41. The question of when learning is most likely to occur is thoroughly explored by M. Leann Brown and Michael Kenney. See their essay "Organizational Learning: Theoretical and Methodological Considerations," in *Organizational Learning*, ed. M. Leann Brown, Michael Kenney, and Michael Zarkin, 1–17.
42. Indeed, some analysts doubt that learning theories can be applied to time periods shorter than a decade because it takes at least this long for policymakers to gain a perspective on the successes and failures of policy. See Paul A. Sabatier, "Policy Change Over a Decade or More," in *Policy Change and Learning*, ed. Sabatier and Jenkins-Smith, 13–39.

CHAPTER 2

ORGANIZATIONAL LEARNING AND REGULATORY AGENCIES

Like other policy learning scholars, organizational learning theorists generally define learning as the use of experience and other sources of knowledge in an attempt to fashion more effective policies. Organizational learning theorists, however, narrowly focus their attention on the learning activities of individual state organizations possessing substantial decision-making authority. More specifically, organizational learning theorists seek to understand how organizations gather, process, store, and disseminate knowledge in a manner that makes it accessible and useful in policy-making situations. In doing so, these theorists augment traditional conceptions of policy learning with additional insights gleaned from organization theory and cybernetic systems theory.[1]

For the purposes of this study, three basic propositions concerning the nature of organizational learning are accepted. First, learning is accomplished when the "memory" of the organization (rather than the beliefs and preferences of individuals) is altered. Organizational memory takes the form of policies, documents, values, and procedures that encode organizational experience. Individuals and groups serve as the agents of organizational learning, acquiring knowledge and disseminating it within the organization. However, as Andrew Bennett notes, "organizational learning is a political as well as a cognitive process … [that] takes place through contestation and bargaining within organizations and between organizations and external political actors." Thus, a policymaking organization is engaged in an ongoing dialog that transcends any particular set of individuals. Over time, changes in organizational personnel and the composition of the external environment alter the way knowledge is gathered and interpreted.[2]

Second, learning processes can be conceptualized as consisting of three "orders," or levels of complexity. First-order learning impacts technical judgments concerning the selection and adjustment of specific policy instruments. Second-order learning is more fundamental in character, involving reflection on the underlying ideas and goals that guide policymaking. Third-order learning is the most complex of the three and occurs when organizational participants demonstrate a willingness to engage in self-reflection and to thus improve the organization's overall capacity to process knowledge and learn effectively.[3]

Third, it is assumed that organizational learning is by no means a foregone conclusion. In some instances, policymakers lack either adequate knowledge or the tools to evaluate knowledge effectively. In other situations, adherence to established ideas and values causes policymakers to filter out potentially useful new knowledge. Sometimes extensive political pressure or legal constraints influence the way knowledge is used or limit the range of acceptable policy choices.[4] Thus, learning is most likely to occur under favorable organizational, institutional, and political circumstances.

These basic parameters of organizational learning theory form the foundation for the remainder of this chapter, which is dedicated to developing key theoretical details. First, a thorough discussion of the three "orders" of learning is provided. Second, the conditions most conducive to effective learning are elaborated. Finally, the theoretical propositions are employed to develop a plan for analyzing the case of early FCC cable television regulation.

First-Order Learning

As previously discussed, first-order learning involves the use of new knowledge to select and adjust policy instruments. When decision makers confront a policy problem that has persisted for quite some time, experience becomes the most immediately relevant form of knowledge. Experience accrues most directly to an organization in the form of "feedback" concerning the success or failure of previously selected policy instruments. As policymaking participants collectively attach meaning to experience, they adjust policy instruments in the manner deemed most appropriate to the attainment of established goals. In instances where policy instruments are no longer considered effective, decision makers select alternative mechanisms from a repertoire of policy choices encoded in the organization's memory. When organizations need to conduct a broader search for new policy options, they may transfer and adapt ideas from other organizations or jurisdictions through "lesson drawing."[5]

First-order learning may be easy to undertake when organizational participants possess extensive, unambiguous experience with a specific policy problem. More frequently, however, experience is limited in scope and its meaning is disputed. This is particularly true in the case of novel policy problems—problems that, because of their uniqueness, present decision makers with a substantial gap between existing knowledge and real-world circumstances. Ideally, novel policy problems could be resolved by likening them to situations an organization has dealt with in the past—a process known as analogical reasoning. When analogies are easily drawn between new policy problems and old ones, organizations feel more comfortable experimenting with their existing repertoire of policy techniques.[6]

Although the utilization of analogies to resolve novel problems might be ideal, the effectiveness of this approach is by no means guaranteed. For one thing, policymaking participants may disagree over the appropriate analogy to apply in a given situation. Furthermore, when unprecedented policy situations are accompanied by high levels of scientific or technical uncertainty, decision makers may feel uncomfortable applying existing policy instruments without further information. In these instances, organizations are likely to accelerate their efforts to acquire policy-relevant knowledge. Through techniques such as in-house research, contracting with outside research organizations, or convening special conferences or proceedings, organizations attempt to reduce technical uncertainty and discern workable policy options.[7]

At the same time, however, high degrees of novelty and technical uncertainty may limit the prospects for adequate knowledge acquisition. Competing research studies based on limited or ambiguous data frequently yield contradictory results. Furthermore, ambiguous scientific findings may fuel political conflict as contending interests advance their own interpretations of the empirical data. Under these circumstances, contestation and dialog become particularly important as decision makers attempt to make sense of multiple, competing interpretations of reality. Amid scientific and political controversy, organizational leaders may work to foster consensus, or they may choose to base decisions on intuition and personal preferences. Whatever the process, the end result is frequently "learning by doing." In essence, the organization arrives at a decision, hoping to build up a body of knowledge and experience that will inform future decision making.[8]

Second-Order Learning

As organizational decision makers struggle to clarify problems and fashion solutions, their actions are frequently informed by preexisting policy beliefs. Although specific conceptions of policy beliefs vary, learning theorists generally characterize them as middle range and policy specific. Thus, whereas policy beliefs may be informed by "bigger" ideas such as ideologies and theories of state-market relations, they mainly include knowledge and techniques arcane to a specific policy area. For this reason, the exact structure of policy beliefs varies across policy domains. Generally speaking, however, policy beliefs combine knowledge obtained from both theory and practice—including ideas, for example, from legal doctrines, scientific theories, research findings, political values, and from the accumulated wisdom of practical problem solving. The end result is a system of ideas and standards that helps policymaking participants identify both the nature of policy problems and the appropriate range of alternatives for resolving them.[9]

Issue areas characterized by technical complexity may spawn multiple competing belief systems over time, as policymaking participants interpret knowledge and experience in different ways. Nevertheless, a single set of policy beliefs may come to dominate for two interrelated reasons. First, a set of policy beliefs is likely to dominate if it is consistently espoused by an influential epistemic community. According to Peter Haas, epistemic communities consist of "professionals with recognized expertise and competence in a particular domain and an authoritative claim to policy-relevant knowledge."[10] As individual members of an epistemic community gain positions of influence in a wide range of venues within a particular policy domain, their beliefs often come to dominate.

Second, a certain set of beliefs may come to dominate a policy area if it becomes embedded within organizational practice. This can occur in a variety of ways—for example, ideas may become encased within the provisions of enabling legislation, in organizational structures, or in accumulated precedents. Taking advantage of civil service protections, members of a dominant epistemic community may hold positions of authority within the organization for extended periods of time, thus deepening the influence of their beliefs.[11] Through a combination of institutions and personnel, key components of a particular belief system become part of the normal organizational discourse.

Second-order learning occurs when a dominant system of policy beliefs is altered over time. Whereas the policy learning literature tends to focus on major alterations in belief systems, relatively minor adjustments occur frequently in response to personnel changes and new knowledge.[12] Members of the organization possess their own mental models through which they categorize events, evaluate data, and fashion responses. Each member's mental model is shaped not only by policy-relevant beliefs but also by personal values and experiences. Thus, even adherents to a dominant policy belief system will interpret its tenets in different ways.[13] Over time, adherents adjust and elaborate their understandings of prevailing ideas in response to new information, experiences, and changes in the external environment. Furthermore, even when policy beliefs are otherwise embedded in organizations, their meaning may be altered or even partially preempted by leaders who adhere to alternative ideas.[14] Ultimately, the exact influence of policy beliefs at any given moment is determined by dialog both within organizations and between organizations and external actors.

Major alterations in belief systems tend to be precipitated by broad changes in the policymaking environment. Elections, changing patterns of interest group power, changes in organizational leadership, and broad societal crises can all stimulate a search for new policy ideas. Nevertheless, even though these circumstances create conditions that facilitate the reexamination of policy beliefs, broad-based change is unlikely unless a consensus emerges within an organization that existing ideas are outmoded. Policy belief systems may break down over time as novel policy problems or "anomalies" emerge that cannot be easily resolved or even fully comprehended through the logic of prevailing ideas and standards. Policymaking participants may first attempt to resolve unprecedented problems by stretching the belief system to accommodate new circumstances; in some cases such efforts only exacerbate the problems or create political conflict. In the wake of repeated policy failures, then, organizations reevaluate old strategies and search for fundamentally new policy ideas and approaches.[15]

Yet, even when conditions are ripe for belief system change, that change is by no means imminent. As previously noted, embedded ideas are difficult to dislodge and it may be more feasible to either change the existing beliefs incrementally or to augment them with new ideas rather than to fully replace them.[16]

In some instances, viable alternative belief systems may not be available at all, forcing decision makers to continue to struggle with existing ideas. But even when new belief systems are available, they cannot take root unless they are advocated by high-profile leaders and new epistemic communities that can work to embed them in organizational practice.[17] When all these factors are taken into consideration, second-order learning of the most fundamental kind is a rare proposition, at best.

THIRD-ORDER LEARNING

When confronted with novel problems, organizations face the possibility that no single available belief system is adequate to address the new challenges. In these instances, the organization may need to step back from all belief systems, knowledge, and experiences and reflect on the policy problem in a more open, detached manner. This is what learning theorists refer to as third-order learning, or "learning to learn." The concept originated in cognitive psychology, where researchers have observed that as individuals acquire experience in solving a particular type of problem, they become more proficient problem solvers. This occurs when individuals deliberately reflect on their actions and thought processes as they engage in problem solving.[18]

According to Argyris and Schon, organizations as well as individuals can learn to learn. Organizations do so when they

> reflect on and inquire into the previous contexts for learning. They reflect on and inquire into previous episodes of organizational learning, or failure to learn. They discover what they did that facilitated or inhibited learning, they invent new strategies for learning, they produce these strategies, and they evaluate and generalize what they have produced.[19]

Thus, learning to learn in this context means that the organization collects, stores, and utilizes knowledge in a manner that is conscious and reflective. With respect to government policymaking organizations, Leann Brown suggested that this process might involve "putting into place organizational structures, programs and processes to draw upon internal and external sources of expertise and undertake … research."[20] Rein and Schon further argued that research and analysis should be reflective—that is, they should inquire into the history and context of hidden policy assumptions and reflect on the validity of those assumptions.[21]

In essence, when organizations learn to learn, they become "learning organizations" in the sense that they are better able to utilize knowledge in the policy process. However, becoming a learning organization is difficult and only likely to occur under the appropriate learning conditions. The conditions most likely to bring about effective learning within regulatory agencies are considered in the following section.

Learning Conditions

Learning theorists generally posit that an organization's learning success is impacted by a combination of endogenous and exogenous conditions.[22] Although theorists point to a range of possible influences, it is argued here that four conditions are crucial to learning in regulatory agencies: pragmatic leadership, organizational capacity, a favorable institutional context, and a favorable political context.

Pragmatic Leadership

Within government agencies, the principal leaders are those in positions of formal decision-making authority, such as agency and bureau heads. Although leaders may achieve results simply through the exercise of formal authority, theorists agree that the most successful leaders are those who build support for favored initiatives through dialog and collaboration with their employees.[23] However, leadership success in and of itself cannot be equated with organizational learning. A leader may successfully implement organizational change without having thoroughly considered the relevant knowledge. The end result may be positive or negative, but in either case the organization cannot necessarily be said to have learned.[24]

Under the right circumstances, nevertheless, leaders may serve as important conduits for organizational learning. Knowledgeable, persuasive leaders may become organizational teachers who embed new policy belief systems within an organization by convincing policymaking participants of the superiority of those beliefs.[25] At the same time, however, leaders are most likely to build learning organizations when the leaders themselves remain open to new information and ideas. Lawrence Dodd argued that government organizations benefit from leaders who "can learn through experience … to reassess their political preconceptions in the light of new conditions and expanding awareness." Dodd noted, too, that leaders are most likely to do this when they take the time to reflect on and reassess their ideas and when they surround themselves with staff who provide honest assessments of policy conditions.[26] Thus, leading a learning organization is a delicate balance: leaders must be prepared to provide strong policy orientations and at the same time remain pragmatic enough to respond to new information and alternate points of view.

Organizational Capacity

Because pragmatic leaders value the opportunity to consider new knowledge, they are likely to desire increased organizational capacity. For the purposes of this study, organizational capacity can be conceptualized as effective organizational structure combined with the ability to engage in policy planning and analysis.[27] Agencies that possess organizational capacity employ a large number of talented professionals capable of analyzing policy alternatives and suggesting appropriate solutions. The mere presence of professional analysts, however, is not sufficient to promote organizational learning. Agencies must be structured in such a way as to allow for the storage of existing information and experience, such that existing data can be combined and integrated over time with new knowledge. As agencies build up a knowledge base, they become better prepared to scrutinize new data and to make independent, informed policy choices. Agencies must also have in place mechanisms that allow decision makers to effectively access policy analysis and advice when it is most needed. For example, centralizing analysis functions in a single office that is well staffed with trusted personnel is one way for an agency to provide for the clear and effective collection, processing, and usage of knowledge.[28]

Institutional Context

Institutional arrangements influence how knowledge and experience are used in a variety of ways. The notion that policy beliefs become embedded in organizational structure and practice is one facet of institutional influence that has already been discussed, but it is not the only factor at work. Two additional institutional considerations shape opportunities for agencies to learn: statutory delegations of authority and decision-making procedures.

Agencies may receive either relatively broad or relatively narrow delegations of statutory authority from the legislative bodies that create them. A broader delegation of authority generally allows an agency to formulate regulations on its own initiative, whereas a narrower delegation may go so far as to delineate the specific types of regulations to be made. Some scholars have expressed concern that broad delegations of authority simply turn agencies into arenas of conflict which come to be dominated by organized interests.[29] However, when agencies confront novel policy problems, broad delegations of authority provide them with the independence to experiment with new policy instruments; narrow delegations of authority do not allow this freedom.[30] For this reason, it is argued here that broad delegations of authority increase the prospects for learning by a regulatory agency.

Decision-making procedures may also promote learning in agencies by forcing them to consider a broad range of available ideas and evidence. In the United States, for instance, regulatory agencies are legally bound to observe standardized procedures, such as those mandated by the APA. Under the APA, agencies must solicit evidence and input from a broad range of interested parties before they can make a regulation. In most instances, agencies meet this requirement by soliciting written public comments on a proposed regulation. Before issuing the final regulation, the agency must demonstrate that it has thoroughly considered the public comments and weighed them along with other available evidence. If procedures are not followed or if evidence is not properly considered, affected parties may seek judicial review. Thus, the agency is held legally responsible for the way it considers evidence. This is not to say that agencies always use administrative procedures in an unbiased, reflective manner, or that better-quality regulations always result from this process. However, the process of collecting comments and data does guarantee that organizational decision makers will have a broad range of evidence in front of them when they make decisions.[31]

Political Context

Regulatory agencies learn best when they are able to act with a relative degree of autonomy with respect to their political environment. Organizational autonomy is by no means assured in any given instance, but policy and administration scholars suggest a number of ways in which it can be brought about. In the early twentieth century, for example, US administrative reformers advocated the independent commission structure for some government agencies in order to grant them institutional autonomy.[32] Scholars today, however, generally agree that autonomy is more an issue of politics than of organizational design. For instance, Daniel Carpenter argued that some US federal agencies have achieved enduring autonomy by building a "bureaucratic reputation," which organizational leaders and their staff create by building networks of support across their policy domain. By employing their professional and political connections, agency personnel build such a high degree of legitimacy that powerful politicians and interest groups are willing to defer to the policy preferences of these experts. According to Carpenter, when bureaucratic reputation exists, bureaucrats are best positioned to learn, experiment, and innovate.[33]

Although his argument is compelling, Carpenter documents only a handful of instances in which agency autonomy was achieved through bureaucratic reputation alone. Taking a different tack, William Gormley posits that bureaucratic autonomy varies depending on the levels of complexity and salience associated with a particular issue area. Policy areas characterized by high levels of technical complexity, for example, require bureaucratic expertise; in such cases, broad delegations of statutory authority are often granted to administrative agencies. Likewise, when high technical complexity is combined with low issue salience, bureaucrats are generally ignored by politicians and courted by interest groups. When interest group demands are diverse and conflicting, bureaucrats gain the opportunity to advocate their preferred policies as they engage in dispute resolution.

It is when complex issues become politically salient, however, that bureaucrats are best positioned to learn. Under these conditions, regulators are forced to strike a balance between their own technical expertise and accountability to external political forces, such as legislative bodies. Legislators may not have the expertise to formulate policy, but they can hold bureaucrats' feet to the fire, forcing them both to take action and to justify their choices. Under circumstances such as these, organizational decision makers are most likely to set aside their preconceptions and consider new ideas and policy options.[34]

Overview of the Study

The purpose of this study is to determine whether organizational learning influenced the FCC's efforts to fashion a cable television policy between 1952 and 1980. Specifically, the study examines the FCC's efforts to evaluate and regulate the economic impact of cable television on over-the-air broadcasting. The FCC dealt with a range of policy questions relating to cable television, but the issue of economic impact was the most vexing and enduring among them.[35] Moreover, resolution of this question required a high degree of technical knowledge—a condition typically associated with policy learning.

There are three additional reasons to suspect that learning played a role in the FCC's decision processes. First, prior studies have suggested that cable television presented regulators with a highly novel, complex policy problem.[36] Second, the FCC formulated and reformulated its policies on multiple occasions during the time period under examination. Although multiple alterations in policy do not equal learning, they do suggest that initial policy formulation efforts were possibly made under conditions of uncertainty and that subsequent experiences demonstrated the need for new choices. Third, the most drastic changes in policy occurred in the late 1970s, a period during which US regulatory policy in general was influenced by new policy beliefs rooted in microeconomic analysis.[37] Taken together, these circumstances provide preliminary support for the thesis that learning influenced the FCC's cable television policies between 1952 and 1980.

Building on the theoretical propositions outlined in this chapter, the following research questions guide the analysis:

1. Did the FCC engage in first-order learning? In other words, was the FCC prompted to refashion policy instruments in light of new knowledge? Assuming that cable television in fact presented the FCC with a novel policy problem, can the use of analogical reasoning or learning by doing (or both) be observed in the early years of policymaking? Are these methods of learning followed by learning from experience later on?
2. Did the FCC engage in second-order learning? That is, did the FCC revisit fundamental policy beliefs and assumptions as they proved unworkable over time? Did the FCC articulate new ways of understanding and acting on cable television policy problems?
3. Did the FCC engage in third-order learning? In other words, did the FCC become a more effective problem-solving organization with respect to cable television? Did the FCC demonstrate signs of increased research and problem-solving capacities and an increased willingness to reflect on the validity of existing policies and assumptions?

4. Did learning conditions advance or inhibit organizational learning? This research question admittedly encompasses a broad range of considerations (discussed earlier in this chapter). In essence, did pragmatic leadership, organizational capacity, and the institutional and political context advance or inhibit learning throughout the period between 1952 and 1980?

Chapters 3–7 present the empirical data. Chapter 3 contains an examination of the ideas and institutional arrangements that were prevalent when cable television was first being developed, and it concludes with a discussion of the FCC's earliest recognition of cable television as a public policy problem. In chapters 4–7, the case study of cable television regulation is developed and presented as a chronology; each chapter covers a major episode of FCC policy formulation. Although the case study is structured to draw attention to theoretically relevant data, the bulk of the analysis is reserved for the concluding chapter.

Endnotes

1. Representative works of particular importance include Barbara Levitt and James G. March, "Organizational Learning," *Annual Review of Sociology* 14 (1988): 319–340; Chris Argyris and Donald A. Schon, *Organizational Learning: A Theory of Action Perspective* (Reading, MA: Addison-Wesley, 1978); Richard L. Daft and Karl E. Weick, "Toward a Model of Organizations as Interpretive Systems," *Academy of Management Review* 9 (1984): 284–295; On the application of organization theory to policy learning, see generally M. Leann Brown, Michael Kenney, and Michael Zarkin, eds., *Organizational Learning in the Global Context* (Aldershot, U.K.: Ashgate, 2006); With respect to cybernetic theory, see John D. Steinbrunner, *The Cybernetic Theory of Decision* (Princeton, NJ: Princeton University Press, 1974); see also Gregory Bateson, *Steps to an Ecology of Mind* (New York: Ballantine Books, 1972).
2. Andrew Bennett, *Condemned to Repetition?: The Rise, Fall, and Reprise of Soviet-Russian Military Interventionism, 1973–1996* (Cambridge, MA: MIT Press, 1999), 99. This basic sentiment is shared by a number of theorists. See M. Leann Brown and Michael Kenney, "Organizational Learning: Theoretical and Methodological Considerations," in *Organizational Learning in the Global Context*, ed. M. Leann Brown, Michael Kenney, and Michael Zarkin, 14 (Aldershot, U.K.: Ashgate, 2006).
3. Learning theorists from a number of traditions agree on this general framework. See Argyris and Schon, *Organizational Learning*, 17–27; Peter Glasbergen, "Learning to Manage the Environment," in *Democracy and the Environment: Problems and Prospects*, ed. William Lafferty and James Meadowcroft, 175–193 (Cheltenham, U.K.: Edward Elgar, 1996); Peter A. Hall, "Policy Paradigms, Social Learning, and the State: The Case of Economic Policymaking in Britain," *Comparative Politics* 25 (1993): 275–296. Hall actually recognizes four "orders" because he considers the selection and adjustment of policy instruments to be two distinct learning activities. Nevertheless, his approach is conceptually similar.

4. On the factors that may impede or facilitate learning, see Brown and Kenney, "Organizational Learning," 7–16; Michael J. Zarkin, "Organizational Learning in Novel Policy Situations: Two Cases of United States Communications Regulation," *Policy Studies* 29 (2008): 89–91.
5. Levitt and March, "Organizational Learning," 320, 330; Richard Rose, *Lesson-Drawing in Public Policy: A Guide to Learning Across Time and Space* (Chatham, NJ: Chatham House, 1993).
6. David Patrick Houghton, "The Role of Analogical Reasoning in Novel Foreign Policy Situations," *British Journal of Political Science* 26 (1996): 523–552; Zarkin, "Organizational Learning," 89.
7. M. Leann Brown, "Scientific Uncertainty and Learning in European Union Environmental Policymaking," *Policy Studies Journal* 28 (2000): 578; Zarkin, "Organizational Learning," 89–90.
8. M. Leann Brown, "Learning and Food Security in the European Union," in *Organizational Learning in the Global Context*, ed. M. Leann Brown, Michael Kenney, and Michael Zarkin, 25–35 (Aldershot, U.K.: Ashgate, 2006); James G. March, Lee Sproull, and Michael Tamuz, "Learning from Samples of One or Fewer," *Organization Science* 2 (1991): 1–3.
9. In developing this general characterization of policy beliefs, I draw liberally from my own work and that of other learning theorists. See particularly Hall, "Policy Paradigms," 279–280; Michael J. Zarkin, "Microeconomic Ideas, Policy Epistemologies, and the Politics of Spectrum Licensing, 1922–1997," *Polity* 38 (2006): 174–193; Lawrence C. Dodd, "Political Learning and Political Change: Understanding Development Across Time," in *The Dynamics of American Politics*, ed. Lawrence C. Dodd and Calvin Jillson, 331–364 (Boulder, CO: Westview Press, 1994).
10. Peter Haas, "Epistemic Communities and International Policy Coordination," *International Organization* 46 (1992): 3.
11. With respect to the embedding of ideas, see Kathryn Sikkink, *Ideas and Institutions: Developmentalism in Brazil and Argentina* (Ithaca, NY: Cornell University Press, 1991); Christopher McGrory Klyza, *Who Controls Public Lands?: Mining, Forestry, and Grazing Policies, 1870–1990* (Chapel Hill: University of North Carolina Press, 1996).

12. Indeed, the literature focuses mainly on the significance of wholesale belief system change rather than adjustment or elaboration of existing beliefs. See, for instance, Hall, "Policy Paradigms," 280; Zarkin, "Microeconomic Ideas," 178.
13. The notion of "mental models" is conceptualized by a number of authors of learning theory literature. For instance, Chris Argyris elaborates on the notion of "theories of action." See Chris Argyris, Robert Putnam, and Diana McLain Smith, *Action Science* (San Francisco: Jossey-Bass, 1985), 81. For a psychological perspective, see Richard Nisbett and Lee Ross, *Human Inference: Strategies and Shortcomings of Social Judgment* (Englewood Cliffs, NJ: Prentice-Hall, 1980), 7, 34, 44–50. In some instances, even entrenched belief systems are interpreted in very different ways. For instance, in the case of British macroeconomic policy in the post–World War II years, Michael Oliver observed that both Labour and Conservative politicians advocated Keynesian policy strategies, though their understandings of what this meant differed significantly. See his *Whatever Happened to Monetarism?: Economic Policymaking and Social Learning in the United Kingdom Since 1979* (Aldershot, U.K.: Ashgate, 1997), 20.
14. This might occur, for instance, when politically appointed agency heads do not adhere to the prevailing ideas.
15. Hall, "Policy Paradigms," 280; Zarkin, "Microeconomic Ideas," 178.
16. Klyza, *Who Controls*, 7.
17. Michael J. Oliver and Hugh Pemberton, "Learning and Change in 20th-Century British Economic Policy," *Governance* 17 (2004): 415–441; Zarkin, "Organizational Learning," 90–91.
18. See Henry Carlton Ellis, *The Transfer of Learning* (New York: Macmillan, 1965); William Kaye Estes, *Learning Theory and Mental Development* (New York: Academic Press, 1970); Gregory Bateson, "Social Planning and the Concept of Deutero-Learning," in *Steps to an Ecology of Mind*, 159–176 (New York: Ballantine, 1972). "Deutero-learning" is Bateson's term for learning to learn. The concept has also influenced studies of professional practice. See Donald Schon, *The Reflective Practitioner: How Professionals Think in Action* (New York: Basic Books, 1983).

19. Argyris and Schon, *Organizational Learning*, 27.
20. Brown, "Learning and Food Security," 36; Brown's conception of learning to learn is similar to that of Jack Levy. See Levy, "Learning and Foreign Policy: Sweeping a Conceptual Minefield," *International Organization* 48 (1994): 286–287.
21. Martin Rein and Donald Schon, "Frame-Reflective Policy Discourse," in *Social Sciences and Modern States: National Experiences and Theoretical Crossroads*, ed. Peter Wagner, Carol Hirschon Weiss, Bjorn Wittrock, and Hellmut Wollman, 267 (New York: Cambridge University Press, 1991).
22. For a review, see Brown and Kenney, "Organizational Learning," 7–16.
23. Leadership studies from a variety of theoretical traditions have reached this conclusion in recent decades. For an overview, see Peter G. Northouse, *Leadership: Theory and Practice*, 2nd ed. (Thousand Oaks, CA: Sage, 2001).
24. See Levy, "Learning and Foreign Policy," 298–300.
25. Marcus Jachtenfuchs, *International Policy-Making as a Learning Process?: The European Union and the Greenhouse Effect* (Aldershot, U.K.: Ashgate, 1996).
26. Lawrence C. Dodd, "Learning to Learn: the Political Mastery of US Senators," *Legislative Studies Section Newsletter* (1992): 10–11; Dodd wrote with US Senators in mind. However, Michael Oliver has found his ideas relevant to the study of governmental leadership in UK economic policymaking in the 1980s. See Oliver, *Whatever Happened to Monetarism?*, 144–146.
27. The importance of organizational capacity in regulatory agencies has been highlighted by numerous authors writing in the state theory tradition. See particularly Stephen Skowronek, *Building a New American State: The Expansion of National Administrative Capacities, 1877–1920* (Cambridge: Cambridge University Press, 1982); Theda Skocpol and Kenneth Finegold, "State Capacity and Economic Intervention in the Early New Deal," *Political Science Quarterly* 97 (1982): 255–278; Marc Allen Eisner, *From Warfare State to Welfare State: World War I, Compensatory State Building, and the Limits of the Modern Order* (University Park: Pennsylvania State University Press, 2000).

28. Organization theorists refer to an organization's ability to effectively utilize and build on a knowledge base as "absorptive capacity." See Wesley M. Cohen and Daniel A. Levinthal, "Absorptive Capacity: A New Perspective on Learning and Innovation," *Administrative Science Quarterly* 35 (1990): 128–152. Organizations possessing absorptive capacity are also comparable to what Daft and Weick refer to as "discovering" organizations. See Daft and Weick, "Toward a Model," 289–290.
29. This argument is most well known as it was made by Theodore Lowi. See *The End of Liberalism*, 2nd ed. (New York: W. W. Norton, 1979), 94–107.
30. See Michael J. Zarkin, *Social Learning and the History of US Telecommunications Policy, 1900–1996: Creating the Telecommunications Act of 1996* (Lewiston, NY: Edwin Mellen Press, 2003).
31. Lloyd S. Etheredge identifies the APA as a potentially important mechanism for learning within government organizations. See his "Government Learning: An Overview," in *The Handbook of Political Behavior*, vol. 2, ed. Samuel S. Long, 125 (New York: Plenum Press, 1981). For an overview of US administrative procedures, see Cornelius M. Kerwin, *Rulemaking: How Government Agencies Write Law and Make Policy*, 3rd ed. (Washington, DC: Congressional Quarterly Press, 2003).
32. Independent commissions are headed by a collegial body whose membership is bipartisan and appointed for overlapping fixed terms of office. By the late 1960s, administrative doctrines influenced by "new left" thinking were arguing that the independent commission design was insufficiently responsive to the public. As a result, many new regulatory agencies like the US Environmental Protection Agency were created as executive agencies or bureaus. By the 1980s and '90s, however, it had become clear that these agencies were easily politicized. For an excellent discussion of these issues, see Marc Allen Eisner, *Regulatory Politics in Transition*, 2nd ed. (Baltimore: Johns Hopkins University Press, 2000).
33. Daniel P. Carpenter, *The Forging of Bureaucratic Autonomy: Reputations, Networks, and Policy Innovation in Executive Agencies, 1862–1928* (Princeton, NJ: Princeton University Press, 2001), 14–36.

34. William T. Gormley, "Regulatory Issue Networks in a Federal System," *Polity* 18 (1986): 595–620.
35. Additional policy questions not covered in this study include ownership issues, rate regulation, and federal-state divisions of authority. The contention that economic impact was a particularly "vexing and enduring" policy question is based in good part on the amount of space these issues received in earlier studies of the FCC and cable regulation. See especially Don R. LeDuc, *Cable Television and the FCC: A Crisis in Media Control* (Philadelphia: Temple University Press, 1973); Stanley M. Besen and Robert W. Crandall, "The Deregulation of Cable Television," *Law and Contemporary Problems* 44 (1981): 77–124.
36. Several studies have asserted that cable television was a new technology that needed to be socially constructed by policymaking participants. See Thomas Streeter, "The Cable Fable Revisited: Discourse, Policy, and the Making of Cable Television," *Critical Studies in Mass Communication* 4 (1987): 174–200; Patrick R. Parsons, "Defining Cable Television: Structuration and Public Policy," *Journal of Communication* 39 (1989): 10–26; Patrick R. Parsons, *Blue Skies: A History of Cable Television* (Philadelphia: Temple University Press, 2008). Don R. LeDuc further asserts that cable television represented an unwelcome technological disruption for an agency that was set up to deal mainly with telephone and broadcast technologies. See LeDuc, *Cable Television and the FCC.*
37. Martha Derthick and Paul J. Quirk, *The Politics of Deregulation* (Washington, DC: Brookings Institution, 1985); see also Zarkin, *Social Learning.*

Chapter 3

Beliefs, Institutions, and the Origins of Communications Regulation

By the time cable television came into existence, communications regulation was already governed by an embedded system of policy beliefs. Telephony and radio broadcasting—then the principal industries of electronic communication—were both legally classified as industries vested with a public interest. The public interest justification for regulation was quite old; having originated in English Common Law, it emerged as a major legal doctrine in the United States during the 1870s. When combined with economic theory in the 1880s, the public interest idea provided justification for the regulation of common carrier industries, such as the railroads. Recognizing basic structural similarities between the railroads and telephony, early twentieth-century regulators extended public interest regulation to the latter industry. In the late 1920s, a distinct version of public interest regulation began to develop for commercial broadcasting. With the passage of the Communications Act of 1934, Congress institutionalized the application of public interest principles to the communications industry. Over the next fifteen years, the FCC would elaborate and institutionalize these principles through its regulatory activities.

Learning theorists predict that when policy beliefs are firmly embedded, efforts to resolve a novel policy problem will begin with analogical reasoning—that is, with a categorization of the new phenomenon in light of existing ideas and precedents. Therefore, when confronted with a new technology like cable television, the FCC might be expected to draw an analogy to either telephony or broadcasting and then to regulate accordingly. In the case of cable television, however, such a move proved difficult because cable television shared similarities with both telephony and broadcasting and neither of the familiar models was entirely adequate to the challenge. Like the telephone, cable television served as a wire-based delivery system for messages on a for-hire basis. Like broadcasting, the messages delivered by cable took the form of commercially developed content obtained from a third party rather than from the customers of the system. Even after the FCC and Congress had concluded that regulation of cable television was in the public interest, it took time to figure out exactly *how* the industry should be regulated.

Thus, before the development of cable regulation can be fully considered, it is first necessary to understand the major policy beliefs and institutional arrangements that were guiding communications regulation at the time of cable's inception. The following discussion explores the core ideas or "models" of communications regulation that had developed prior to cable's appearance, the legal and organizational structures in which these models were embedded, and their application to cable television in the earliest days of that medium.

The Common Carrier–Public Utility Model

The core principles of modern public utility regulation were derived during the late nineteenth century from a combination of legal precedent and economic theory. In legal terms, public utilities were accorded a status similar to that given "common callings" during the Middle Ages. A common calling was broadly defined in common law as an industry or profession vested with a public interest. In essence, common callings were enterprises of such broad social importance that their services needed to be made available on a non-discriminatory basis. In order to provide such extensive access, an exceptional degree of governmental control was allowed over these industries, including the prescription of consumer rates.[1]

From a regulatory standpoint, common carriers—business enterprises that transported the goods of others on a fee-for-service basis—were a particularly important category of common callings. Although mail delivery and wagon operators had been the major examples of such enterprises, by the 1870s the railroads were emerging as the common carrier of the industrial era. Manufacturers and farmers found the railroads indispensible for the transportation of goods, but the industry was prone to collusive business practices that resulted in excessive rates. Responding to pressure from disaffected groups, several midwestern states passed maximum-rate laws and set up regulatory commissions.[2]

During the 1880s, Progressive Era economic theorists argued that government regulation was most appropriate when common carriers were also "natural monopolies": enterprises with start-up costs so high that they could only be recovered through a high volume of output over time. Competition proved prohibitive in these industries precisely because competition distributed output and made it difficult for firms to remain profitable.[3] Indeed, the railroads were a prime example of how high start-up costs and competition could not coexist. Once the expenditures to lay tracks and purchase cars had been made, railroads had to run at maximum capacity in order to survive. However, with open competition, railroad firms could not simultaneously maintain the necessary rate levels and still attract enough customers to remain profitable. As a result, railroads were forced to engage in price fixing and eventually consolidate into a smaller number of firms. These realities quickly led Progressives to call for governmental control over the railroads, through either public ownership or the regulation of private enterprise.[4]

Given the general preference for private enterprise that dominated late nineteenth-century legal thought, policymakers were naturally drawn to regulation rather than public ownership. However, regulating an industry as complex as the railroads presented a novel challenge for a governmental system with little institutionalized expertise or administrative machinery.[5] By the end of the 1880s, Progressive Era theorists and politicians had settled on a remedy for this deficiency: the establishment of independent regulatory commissions. In addition to the fact that commissions represented a familiar organizational model,[6] Progressives preferred them because they were intentionally designed to insulate the technical business of regulation from politics. Created as bipartisan bodies whose members served for staggered, fixed terms of office, commissions limited the ability of any single political party or administration to unilaterally control regulatory policy. This insulation from politics also enabled the legislatures to entrust commissioners and their staff with broad delegations of rulemaking authority.[7] In the case of railroad commissions, like the US Interstate Commerce Commission (ICC), the rulemaking responsibilities came to include the power to set maximum rates and to determine market entry.[8]

By the turn of the twentieth century, the common carrier–public utility regulatory model was being extended to the telephone industry. Indeed, the telephone shared a number of characteristics with existing common carriers: like the mail and the telegraph, the telephone served as a neutral conduit for the delivery of messages; like the railroads, the telephone bore the characteristics of a natural monopoly.[9] By 1910, more than twenty states had given regulatory jurisdiction over local telephone service to their public utility commissions.[10] That same year, Congress gave the ICC authority to regulate long-distance telephony using the existing public utility framework that already applied to the railroads.[11] Although some modifications were made over the years, by the time the FCC inherited authority over long-distance telephony in 1934, the common carrier–public utility model of regulation was the established regulatory philosophy in that field.

The Broadcasting–Public Interest Model

Unlike telephony, radio broadcasting could not easily be placed within the common carrier–public utility model. For one thing, broadcasters originated the messages they delivered; they did not simply carry the messages of others on a for-hire basis. In this regard, broadcasters more closely resembled newspapers than telephones. Moreover, broadcasting was not a natural monopoly; economies of scale were low enough that multiple broadcasters could easily compete within a single listening area. Instead, by the early 1920s, calls for regulation of the industry were motivated by concern over the growing number of broadcasters who were seeking access to a finite number of electromagnetic frequencies.

Recognizing that some system of priorities was needed to guide the licensing of the limited broadcasting frequencies, the then secretary of commerce Herbert Hoover called a series of four radio conferences between 1922 and 1925, in order to solicit the opinions of experts and entrepreneurs from both industry and government.[12] From quite early on, Hoover expressed the belief that broadcasting should serve the public interest and that some type of government involvement might facilitate this goal.[13] For Hoover, the public interest was served when the needs of "the great body of the listening public" were satisfied.[14] In principle, this conception of the public interest was not very different from the public utility model, in which non-discriminatory, equitable service was the standard. In practice, however, the public interest inherent in broadcasting meant that licenses should be reserved for those stations that had the technical capacity to reach a broad audience. In this context, competition—rather than monopoly regulation—would be the impetus driving broadcasters to expand their audience appeal.

Because Hoover had limited authority as secretary of commerce, he was unable to fully implement his vision of the public interest. When Congress passed the Radio Act in 1927, however, the public interest concept was given a prominent place in broadcast station licensing. This piece of legislation authorized the newly created Federal Radio Commission (FRC) to distribute and renew licenses when such licensure was found to be in the "public interest, convenience, and necessity." Although the law provided no definition of the phrase, the FRC interpreted it in terms quite similar to those envisioned by Secretary Hoover years earlier. For the FRC, the public interest, convenience, and necessity meant "good service" attained through a "modern, efficient apparatus."[15] Such technically superior service would allow broadcasters to realize the longstanding public interest principle of non-discrimination which, in the field of broadcasting, meant that "the entire listening public of a station … is entitled to service … [and] the tastes, needs, and desires of all substantial groups … should be met."[16]

The FRC was in existence for only seven years, and the agency never had the chance to fully develop a set of licensing criteria rooted in its definition of the public interest. That task, however, would be undertaken following passage of the Communications Act of 1934, when authority to regulate radio was transferred to the FCC.

The Communications Act of 1934

Although the Communications Act of 1934 added some new features to both the common carrier and broadcast regulatory models, the legislation's main purpose was to consolidate federal regulatory authority and place it in the hands of the FCC. Drawing on the longstanding non-discrimination principle associated with public interest regulation, the Communications Act's primary goal was "to make available, so far as possible, to all people of the United States a rapid, efficient, nation-wide and world-wide wire and radio communications service with adequate facilities at reasonable charges."[17] In addition to granting this broad mandate, the language and structure of the law reinforced the distinct nature of the two existing models of communications regulation.

Title II of the Communication Act provides the basic regulatory framework for common carriers, which are defined as "any person engaged as a common carrier for hire, in interstate or foreign communication by wire or radio, or in interstate or foreign radio transmission of energy." This definition is inclusive, but it is qualified by one carefully stated caveat: "a person engaged in radio broadcasting shall not … be deemed a common carrier." Under Title II, it is unlawful for common carriers to engage in "unjust or unreasonable discrimination in charges … or services," and the FCC is empowered to investigate all new common carrier rates and charges and to prescribe "just and reasonable" rates when necessary. Finally, Title II gives the commission authority to control entry into common carrier markets by approving the extension of lines.

Title III contains the regulatory provisions for radio communication, which is defined as "the transmission by radio of writing, signs, signals, pictures, and sounds of all kinds, including all instrumentalities, facilities, apparatus, and services incidental to such transmission." The broadcast licensing provisions are largely carryovers from the earlier Radio Act, including the basic requirement that license awards and renewals be based on public interest considerations. Additionally, the conditional nature of a broadcast license is reinforced by provisions in Title III that limit the term of a license to three years and empower the FCC to revoke a license if a broadcaster fails to "observe any of the restrictions and conditions of this Act or of any regulation of the Commission."

The Communications Act was passed at the behest of President Roosevelt, who stated in a message to Congress that consolidation of regulatory authority would bring "clarity and effectiveness" to the communications field.[18] However, achieving the clarity and effectiveness FDR sought proved a difficult task for a single agency; the FCC had in effect been charged with regulating two distinct industries, each of which involved quite different technical considerations. Common carrier regulation principally consisted of rate setting, which required the expertise of accountants and economists. Guaranteeing "good service" in broadcasting involved setting technical standards; this task, in turn, required the specialized knowledge of radio engineers. Thus, it seemed logical that the FCC would organize itself into more specialized operating units in order to better accommodate these technical distinctions within the communications industry.

Indeed, Congress seemed to have anticipated this problem; included in the Communications Act was a provision authorizing the FCC to segment its seven commissioners into divisions organized around specialized work responsibilities. The FCC had experimented with divisions in the 1930s, but it ultimately chose to deal with the common carrier–broadcasting dichotomy through a more specialized form of staff organization. By 1950, the commission had segmented its staff into two specialized offices: the Common Carrier Bureau and the Broadcast Bureau. Each bureau was headed by a bureau chief who reported directly to the commission.[19] Although the commission itself retained final approval over all agency actions, the technical expertise of the bureaus gave them tremendous influence over major decisions within their respective jurisdictions. Thus, over time, the bureaus and their chiefs came to be gatekeepers of expert knowledge and, in some instances, strong advocates for their own preferred policy positions.[20]

The FCC and Public Interest Broadcasting, 1940–1960

During the three decades following the FCC's creation, the agency used its statutory discretion to elaborate both the common carrier and the broadcasting regulatory models. The FCC's actions in the area of broadcasting, however, are particularly relevant to its early efforts to regulate cable television. By the 1960s, the FCC had come to view cable as a technology ancillary to broadcasting. In this regard, cable television was only useful to the extent that it helped advance, or at least did not undermine, the public interest mandate of broadcasting. Thus, it is necessary to consider how the FCC defined the public interest in broadcasting during the period under consideration.

By the end of the 1940s, the FCC had taken the position that at license renewal time, broadcasters should be held responsible for demonstrating that the content of their programming served the public interest. Recall that in the late 1920s, the FRC had defined the public interest to include a well-rounded body of programming that served the broad needs of the listening public. The FRC, however, had believed that free-market competition would lead to well-rounded programming once technically superior stations were licensed. By contrast, the FCC viewed the public interest as a set of social goals to be achieved through government regulation. FCC policies during this time were based on the premise that broadcasting held great potential not only as an entertainment medium but also as a tool for enlivening civic debate and democratic participation. These benefits could not be realized, however, if broadcast content decisions were driven solely by the profit motive or, worse, by political interests of station owners. Marketplace competition remained an important mechanism for bringing well-rounded programming to the public, but the FCC supplemented competition with policy statements and formal rules designed to guide broadcasters in their selection of programming. In its development of these policies, the FCC was guided by two interrelated goals: ensuring a diversity of voices and supporting localism.

In principle, the aim of ensuring a diversity of voices was no different than the FRC's goal of well-rounded programming. Beginning in the early 1940s, the FCC took steps through policy statements to ensure that diverse voices were heard on the air. In the "Mayflower Doctrine," for example, the agency addressed the question of whether broadcasters should be allowed to editorialize. In this instance, the FCC concluded that "radio can serve as an instrument of democracy only when devoted to the communication of information and the exchange of ideas fairly and objectively presented. A truly free radio cannot be used to advocate the cause of the licensee."[21] Eight years later, the FCC revised its views slightly in the "Fairness Doctrine," which stated that broadcasters could editorialize as long as they presented contrasting views. In issuing this policy, the FCC made clear its position that the public should necessarily have "a reasonable opportunity to hear different opposing positions on the public issues of interest and importance in the community."[22]

Ensuring a diversity of voices meant that broadcasters should present content of interest to both national and local audiences. This was, in part, the logic behind the FCC's 1941 chain broadcasting regulations, which prevented networks from interfering excessively with the local affiliates' right to choose the programming they deemed most relevant to their listening audience. To this end, affiliates could not be required to carry only network programming or even specific portions of network programming.[23] Through these regulations, the FCC essentially maintained that a balance between programming of national and local interest was necessary in order to "bring to the people of the nation a diversified program service."[24]

In many respects, the chain broadcasting regulations represent the first major FCC policy action to explicitly support localism—the notion that broadcast licensees must remain responsive to the listening tastes and needs of the local community. Subsequent policy statements expanded the idea of localism by suggesting that stations should also serve as an outlet for local self-expression. For instance, in a 1946 policy statement entitled *Public Service Responsibility of Broadcast Licensees*, the FCC suggested that well-balanced programming should include local live programming.[25] In 1960 the FCC further formalized its encouragement of localism by requiring licensees to undertake a "diligent, positive, and continuing effort ... to discover and fulfill the tastes, needs, and desires" of their local audiences.[26] In conjunction with this mandate, broadcasters were required at license renewal time to demonstrate that they were providing adequate opportunities for local self-expression and for the development of local talent.[27]

Television and the Public Interest

Prior to 1950, virtually all of the FCC's public interest guidelines had been written with radio in mind. Television did not exist as a commercial enterprise in the United States until 1945, and it remained in its infancy until the 1950s. Nevertheless, in their definition of radio communication as the "transmission ... of ... pictures ...and sounds," the authors of the Communications Act had anticipated that television would someday be governed by Title III licensing criteria. Thus, in the late 1940s, when the FCC took up the task of planning a national distribution of television frequencies, it seemed only natural that the same public interest goals which applied to radio would also be applicable to the new medium.

In 1952 the national television frequency plan was completed and went into effect with the issuance of the FCC's *Sixth Report and Order*. In developing the plan, the FCC was guided by five hierarchically ordered goals:

1. To provide at least one television service to all parts of the United States.
2. To provide each community with at least one television broadcast station.
3. To provide a choice of at least two television services to all parts of the United States.

4. To provide each community with at least two television broadcast stations.
5. Any channels which remain unassigned under the foregoing priorities will be assigned to the various communities depending on the size of the population of each community, the geographical location of such community, and the number of television services available to such community from television stations located in other communities.[28]

A close examination of these goals reveals that they reflect established public interest priorities. The most basic goal in the hierarchy is the longstanding public interest goal of non-discrimination—ensuring that the entire population had access to television service. This could be accomplished by importing a television signal from a distant community if necessary, but the second goal suggests that localism was given high priority.

Furthermore, localism was given higher priority than competition. For instance, the second and third goals, taken together, suggest that it would be preferable for a community to have one local television service of its own rather than any number of competing stations imported from other communities. Thus, it seems that the FCC gave priority to localism over program diversity. Recall, however, that by the 1940s, the FCC's public interest goals were based on the premise that regulation could ensure well-rounded programming. From the FCC's perspective, a single, properly regulated local station might do a better overall job of serving the public interest than two imported signals would. Although competition might have been the long-term goal, then, it was not the most immediate priority.

To advance its television goals, the FCC engineered a table of assignments through which it allotted over two thousand channels to nearly thirteen hundred communities throughout the

United States and its territories.[29] In setting up this table, the FCC chose to use a combination of frequencies from both the Very High Frequency (VHF) and Ultra High Frequency (UHF) spectrum bands to deliver television. Twelve channels of VHF spectrum had been in use since 1945 and were considered technically superior to the UHF frequencies. At the same time, however, VHF frequencies were scarce and could not provide the number of channel assignments necessary to build a nationwide television system with competing outlets. Nor, for that matter, would VHF make it possible for every community to have its own local television outlet. Therefore, the FCC chose to keep the existing VHF frequencies and to supplement them with enough UHF assignments to fill in the gaps.[30]

The FCC also made the controversial decision to "intermix" VHF and UHF assignments within the same communities. Two of the major television networks—CBS and DuMont—argued against intermixture on the grounds that UHF was an inferior television delivery system that would be unable to compete with VHF stations for advertising dollars. FCC engineers, however, supported intermixture, believing it to be both technically and economically efficient in the long run.[31] For the FCC, legal considerations also played into the decision-making process; because the Communications Act mandated a "fair, efficient, and equitable" distribution of frequencies among the states and communities, the FCC did not believe it could deny some areas access to the technically superior VHF frequencies. Thus, a policy of intermixture was adopted as a way to make sure that moderate-sized communities had reasonable access to VHF.[32]

Within just a few years of the television plan's implementation, however, it became apparent that UHF was not thriving under the policy of intermixture. In particular, it came to the FCC's attention that UHF transmitting and receiving equipment was inferior and that few television manufacturers were choosing to make UHF-ready sets. Because of these deficiencies, advertisers were unlikely to buy time on UHF stations, particularly when they had the choice of advertising on a VHF station in the same area. By 1955, therefore, the FCC was weighing alternative strategies for jump-starting UHF growth—proposals included revising the table of assignments so as to eliminate intermixture or even moving all television stations to the UHF band.[33]

Following an administrative inquiry, the FCC concluded that all-UHF television was the best long-term mechanism for spurring television growth. Accomplishing this transition, however, would require both re-engineering the table of assignments and gaining the political support of the broadcast industry—two tasks which the FCC was not anxious to undertake. Therefore, the FCC set no timetable for accomplishing the shift to all-UHF; the agency simply committed to further study of the matter.[34] Meanwhile, throughout the remainder of the 1950s, UHF continued to flounder.[35]

CATV Enters the Picture

The development of cable television was a direct response to the slow growth of broadcast television. Even as the first commercial stations went on the air in the mid-1940s, remote communities were already erecting master antennas to pull in television signals from distant cities. By the end of the decade, master antennas had morphed into the first cable television systems.[36] Community Antenna Television (CATV), as it was known for the first decade and a half of its existence, was initially little more than commercially operated systems of hilltop or rooftop antennas that brought service into homes through coaxial cable connections. Thirteen commercial CATV systems were in operation by the end of 1950; two years later, seventy systems were in existence, serving roughly fourteen thousand subscribers.[37]

The FCC began considering the policy implications of CATV in 1951when system operator J. E. Belknap and Associates asked the FCC to license operation of a microwave relay system capable of importing television signals from Memphis, Tennessee, to customers in Missouri and southern Illinois. Soon afterward, the FCC asked its staff to investigate whether CATV might fall within the agency's regulatory jurisdiction. In the spring of 1952, a staff memorandum was issued containing recommendations that the FCC explore the possibility of regulating CATV under the terms of the Communications Act.[38]

Officially endorsed by the FCC's general counsel, the Common Carrier Bureau chief, and the Broadcast Bureau chief, the memo contains a detailed legal analysis in which the prospects of regulating CATV as both a radio service and a common carrier are considered. CATV was an unlikely candidate for Title III radio regulation unless it could be shown to be among the "instrumentalities, apparatus, and services ... incidental" to broadcasting. Though a case might be made that CATV met this criterion, the memo declares that such a claim would involve a "highly technical argument" that was not supported by the general intent of the law.[39]

Common carrier regulation under Title II seemed a more promising avenue if CATV's status as a carrier in the "ordinary" legal sense could be demonstrated. According to the staff memo, FCC common carrier regulation could only be applied if CATV was (1) interstate in character, (2) a communications service, (3) available to the public, and (4) for hire. CATV indisputably met the second, third, and fourth criteria, but whether it was an interstate business enterprise was debatable. On the one hand, precedents drawn from railroad law suggested that carriers could be considered interstate only when they transported goods across state lines. Since few cable systems actually transported television signals across state lines, CATV business activities could not be classified as interstate by this standard.[40] On the other hand, the memo points out that both the Communications Act and FCC precedent at the time defined all wire-based transmission of radio communication as interstate commerce.[41] Thus, because CATV systems intercepted radio transmissions and delivered them over wires, CATV operators might be considered interstate common carriers.

Admittedly, CATV was functionally different from common carriers like the telephone, which acted as a two-way transmission medium. The FCC memo asserts, however, that CATV was no different than one-way "reception services," which were generally classified as common carriers by the FCC. One such service that bore similarities to CATV was RCA's "blind-reception service," through which program material was intercepted from overseas and distributed to subscribers for purposes of monitoring or rebroadcast.[42] Because CATV was simply intercepting broadcast signals and delivering them to subscribers without a return channel, it was structurally similar to this and other reception services.

Although the memo does not conclude with any specific policy recommendations, it urges the FCC to consider several legal questions, including whether it was in the public interest to grant microwave licenses of the kind requested by J. E. Belknap and whether CATV systems were violating the property rights of broadcasters by intercepting and retransmitting their signals without permission.[43] In the meantime, the authors of the memo recommended no further action on the Belknap application or on any other matter that might either advance or inhibit CATV. The authors of the memo seemed to anticipate a multi-year process that would involve further study followed by the possible formulation and eventual adoption of regulations. It seems that the FCC personnel believed that the future of CATV as a component of the national television system would come into focus as information and experience accrued throughout this process.

CONCLUSION

By the 1950s, US communications regulation had come to be influenced by a set of institutionalized policy beliefs rooted in legal doctrine, economic theory, and practical experience. Both telephony and broadcasting were classified as public interest enterprises, and as such they were subject to distinct types of regulation tailored to the technical and economic realities of each industry. These policy beliefs were embedded in both the statutory framework that guided communications regulation and the precedents created by the FCC over the course of more than a decade. Although television did not exist as a commercial enterprise when the Communications Act was written, Congress had anticipated its arrival and developed a definition of "radio broadcasting" expansive enough to include the new technology.

Thus, when the FCC began formulating policy for television in the early 1950s, the agency was operating under the assumption that precedents taken from radio broadcasting could easily be applied to the new medium. As a result, FCC regulators chose to make localism and other existing broadcast policy objectives prominent considerations in their planning of the national television system. The slow development of television during the 1950s, however, not only made implementing localism a challenge but also produced an unintended consequence: the development of a fledgling industry called community antenna television. With CATV still in its infancy, the FCC began the process of figuring out how the new industry fit into the overall structure of the national communications system and whether some form of government regulation was needed.

In this respect, the 1952 FCC staff memo is best understood as an exercise in analogical reasoning in which the FCC attempted to categorize CATV in terms of existing ideas and precedents. The legal issues raised in the memo might have motivated the FCC to exert regulatory control over CATV, but the agency declined to pursue the matter, most likely because it faced more pressing priorities at the time: in 1952 the FCC was working furiously to finalize the *Sixth Report and Order* and to bring broadcast television to the entire nation. CATV might have fit the model of one or another type of public interest industry, but at the time it was a marginal technology with an uncertain future.

Nevertheless, the memo anticipates two major issues that would dominate CATV policy discussions for more than two decades. The first is the issue of "distant-signal importation" posed by the J. E. Belknap application for microwave relay service. Given the FCC's commitment to localism in television, what were the implications of allowing a cable system to import broadcast signals into communities as a substitute for local stations? The second issue is the question of the broadcasters' property rights with respect to their programming. Under existing copyright laws, could broadcasters exert control over the interception and use of their programming by CATV systems?

These issues are posed as rhetorical questions here precisely because in 1952 they were little more than hypothetical matters. Throughout the 1950s, however, these issues gradually moved from the realm of the hypothetical to the realm of reality as broadcast television floundered and CATV continued to thrive.

ENDNOTES

1. Although it originated under English Common Law, the common calling doctrine was brought into the US legal system through the case of *Munn v. Illinois* 94 US 113 (1877). On the history of the common calling doctrine, see Martin Glaeser, *Outlines of Public Utility Regulation* (New York: Macmillan, 1927), 162–172.
2. George Miller, *Railroads and the Granger Laws* (Madison: University of Wisconsin Press, 1971); Solon Buck, *The Granger Movement: A Study of Agricultural Organization and its Political, Economic, and Social Manifestations 1870–1880* (Cambridge, MA: Harvard University Press, 1913).
3. The most important proponents of this argument were Henry C. Adams and Richard T. Ely. For an overview of their ideas, see Thomas Hazlett, "The Curious Evolution of Natural Monopoly Theory," in *Unnatural Monopolies: The Case for Deregulating Public Utilities*, ed. Robert W. Poole, Jr., 8–11 (Lexington, MA: Lexington Books, 1985).
4. Among Progressives, Charles Francis Adams provided perhaps the most prominent and cogent analysis of the railroad problem. For an overview of his ideas, see Thomas K. McCraw, *Prophets of Regulation* (Cambridge, MA: Belknap Press, 1984), 7–14.
5. Stephen Skowronek, *Building a New American State: The Expansion of National Administrative Capacities, 1877–1920* (Cambridge: Cambridge University Press, 1982), 19–35.
6. Ibid., 124, 134.
7. Marc Allen Eisner, *Regulatory Politics in Transition*, 2nd ed. (Baltimore: Johns Hopkins University Press, 2000), 42–43.
8. Ibid., 49–58.
9. Ithiel de Sola Pool, *Technologies of Freedom* (Cambridge, MA: Belknap Press, 1983), 100–101.
10. Indeed, the monopoly status of the telephone industry was accepted during the first years of the twentieth century, after more than a decade of experimentation with competition. See Alan Stone, *Public Service Liberalism: Telecommunications and Transitions in*

Public Policy (Princeton, NJ: Princeton University Press, 1991), 143–148, 160–164.

11. For a good discussion of federal regulation of telephony prior to 1934, see Hamilton G. Loeb, "The Communications Act Policy Toward Competition: A Failure to Communicate," *Duke Law Journal* no. 1 (1978): 5–16.
12. Louise Benjamin, "Working it out Together: Radio Policy from Hoover to the Radio Act of 1927," *Journal of Broadcasting and Electronic Media* 42 (1998): 221–236.
13. In his speech at the first radio conference, Hoover noted that "it is of primary public interest to say who is to do the broadcasting, under what circumstances, and with what type of material." Quoted in Marvin R. Bensman, *The Beginning of Broadcast Regulation in the Twentieth Century* (Jefferson, NC: McFarland and Co., 2000), 50.
14. Quoted in Willard D. Rowland, Jr., "The Meaning of 'Public Interest' in Communications Policy—Part II: Its Implementation in Early Broadcast Law and Regulation," *Communications Law and Policy* 2 (1997): 371.
15. US Federal Radio Commission, *Third Annual Report to Congress* (Washington, DC: GPO, 1929), 33.
16. Ibid., 34.
17. Language quoted from the Communications Act in this section is taken from a reprinted version contained in Max D. Paglin, ed., *A Legislative History of the Communications Act of 1934* (New York: Oxford University Press, 1989), 921–966.
18. Senate, *A Message from the President of the United States Recommending that Congress Create a New Agency to be Known as the Federal Communications Commission*, 73rd Cong., 2nd Sess., 1934, S. Doc. 144.
19. Michael Zarkin, "The Evolution of Bureaucratic Design: The Development of the Federal Communications Regulatory Bureaucracy, 1926–1952" (Paper presented at the annual meeting of the Western Political Science Association, Albuquerque, NM, March 2006), 17–23.
20. As one former bureau chief noted, the position became a "power center in shaping regulatory programs." See Fred W. Henck and

Bernard Strassburg, *The Slippery Slope: The Long Road to the Breakup of AT&T* (Westport, CT: Greenwood Press, 1988), 20.
21. *In the Matter of Mayflower Broadcasting Corp.*, 8 FCC 339, 341 (1941).
22. *In the Matter of Editorializing by Broadcast Licensees*, 13 FCC 1246 (1949).
23. Herbert H. Rosenberg, "Program Content—A Criterion of Public Interest in FCC Licensing," *Western Political Quarterly* 2 (1949): 382.
24. Ibid., 383.
25. This document is more commonly known to broadcasting scholars as the "Blue Book." An edited version can be found in Frank J. Kahn, ed., *Documents of American Broadcasting*, 4th ed. (Englewood Cliffs, NJ: Prentice-Hall, 1984), 148–163.
26. *En Banc Programming Inquiry*, 44 FCC 2303, 2312 (1960). This process was more commonly referred to as "ascertainment."
27. Ibid., 2314.
28. *Sixth Report and Order*, 41 FCC 148, 167 (1952).
29. US Federal Communications Commission, *Eighteenth Annual Report to Congress* (Washington, DC: GPO, 1952), 109.
30. *Sixth Report and Order*, 168.
31. Hugh R. Slotten, *Radio and Television Regulation: Broadcast Technology in the United States, 1920–1960* (Baltimore: Johns Hopkins University Press, 2000), 179–181.
32. *Sixth Report and Order*, 168–171.
33. US Federal Communications Commission, *Twenty-Second Annual Report to Congress* (Washington, DC: GPO, 1956), 94–95.
34. US Federal Communications Commission, Report and Order, 25 June 1956, Docket 11532 (FCC Docketed Case Files, RG 173; Washington, DC: National Archives), 7–11.
35. US Federal Communications Commission, *Twenty-Fifth Annual Report to Congress* (Washington, DC: GPO, 1959), 53.
36. Patrick R. Parsons, *Blue Skies: A History of Cable Television* (Philadelphia: Temple University Press, 2008), 59–71.
37. Don R. LeDuc, *Cable Television and the FCC: A Crisis in Media Control* (Philadelphia: Temple University Press, 1973), 68.
38. Parsons, *Blue Skies*, 122–128.

39. FCC Staff, Interoffice Memorandum, 25 March 1952, Reprinted in Senate Committee on Interstate and Foreign Commerce, *Television Inquiry—Part VI: Hearings before the Committee on Interstate and Foreign Commerce*, 85th Cong., 2nd Sess., 1958, 3494.
40. Ibid., 3495.
41. Ibid., 3496.
42. Ibid., 3496–3498.
43. Ibid., 3499.

Chapter 4

Making Sense of Economic Impacts, 1956–1960

The 1950s constituted a decade of rapid growth in CATV. Whereas in 1952 there had been fourteen systems serving a few thousand homes, by the end of the decade there were over six hundred systems serving roughly seven hundred fifty thousand homes.[1] The growth in CATV during this time was certainly aided by the absence of any FCC regulation. Following the 1952 staff memo, the FCC did not explore the prospect of regulation again for another four years. If anything, FCC actions during that time had the effect of promoting CATV development. The agency agreed to approve the application of J. E. Belknap and Associates to provide common carrier microwave relay service to CATV systems in 1954.[2] Over the next several years, The FCC approved the applications of numerous similar microwave common carriers, thereby enabling CATV systems to import television signals from distant locations. This single policy change enabled the rapid and continuous growth of CATV well into the 1960s.

While CATV was growing, however, UHF television was still floundering. By 1959, more than 90 of the 167 UHF stations that had been in operation since 1952 had gone off the air.[3] Although the problems associated with UHF broadcasting certainly predated the growth of CATV, by the mid-1950s some station owners were complaining that the new medium was intensifying their struggle to remain financially viable. These station owners spearheaded a renewed call for the FCC to exert regulatory jurisdiction over CATV, though to no avail; by the late 1950s, the FCC was headed by the Republican chairman John Doerfer, a reluctant regulator who preferred a narrow interpretation of statutory authority. Without a specific grant of authority, the Doerfer commission was unwilling to impose regulation on CATV.

Although the Doerfer commission did ask Congress for the authority to regulate CATV, legislation was stalled by disagreements between broadcasters and CATV operators. Although these groups undoubtedly held political sway over many members of Congress, the power of uncertainty proved to be equally compelling. The causes of UHF stagnation were poorly understood in 1959, and CATV was only one of many potential factors. In an environment characterized by uncertainty, the personal beliefs of regulators and the political preferences of interest groups proved to be stronger influences than technical knowledge.

The Doerfer Commission

Having conducted a thorough study of all the FCC commissioners who served between 1934 and 1961, Lawrence W. Lichty concluded that "personal experience, education, occupational background, and governmental philosophy … directly influence[d] the direction and emphasis of the agency's policies."[4] Lichty's observation certainly describes the period in FCC history from 1957 to 1960, the period during which the agency was headed by Chairman John Doerfer. A lawyer by training, Doerfer joined the FCC in 1953 after a stint as chairman of the Wisconsin Public Utilities Commission. Doerfer was elevated to the position of FCC chairman in 1957 when George C. McConnaughey resigned.[5]

Doerfer favored a free-market philosophy unless the FCC was legally required to regulate; for this reason, he was hesitant to exert regulatory control over CATV simply to protect broadcasters.[6] When questioned before Congress as to why the FCC was not regulating CATV, Doerfer stated that he "could no more protect a television station from a community antenna service than … protect it from a local drive-in movie house or whatever other form of attraction might develop within the service area."[7]

Doerfer likewise favored a narrow reading of the FCC's statutory authority. In 1956 when the FCC tentatively voted to move all television assignments to the UHF band, Doerfer dissented; he believed that the decision violated the language of the Communications Act. To Doerfer, the legal mandate to provide a "fair, efficient, and equitable" distribution of frequencies meant that the FCC should first and foremost ensure that every community had access to some form of television service. By mandating all-UHF, however, Doerfer believed the agency had expressed greater concern for protecting localism and ensuring the economic viability of UHF stations. At a time when some areas of the country still lacked access to television service of any kind, Doerfer considered this unacceptable.[8]

During Doerfer's tenure as chairman, the FCC as a whole was generally conservative, and commissioners of both parties supported the chairman's free-market positions. Robert T. Bartley, for example, favored minimizing government interference in communications markets whenever possible.[9] Likewise, Commissioner Robert E. Lee summed up his own regulatory philosophy by stating that "the public interest is well served when government restraints are kept to a minimum."[10] During his twenty-odd years at the FCC, Commissioner Rosel Hyde took a skeptical view of any regulation that was not clearly in the public interest.[11] T. A. M. Craven, too, preferred a narrow reading of statutory authority and concluded that FCC regulation was "basically unsound" during his many years as both a staff member and commissioner.[12] Although these individuals certainly did not agree on every issue, their respective experiences and regulatory philosophies, when combined, led them to the conclusion that the FCC lacked the clear authority to regulate CATV.

THE *FRONTIER BROADCASTING* CASE, 1956–1958

Frontier Broadcasting was the first formal proceeding in which the FCC considered the matter of CATV regulatory jurisdiction. The proceeding was initiated in 1956 when thirteen broadcasters filed a complaint against 288 CATV systems in thirty-six states.[13] The complainants asked the FCC to assert regulatory jurisdiction over CATV in order to prevent economic harm to broadcasters. Broadcasters alleged that CATV was frustrating the objectives of the *Sixth Report and Order*, particularly the goals of providing one television service to every part of the country and at least one in each community. By bringing in signals from other communities, CATV was diminishing advertising revenue to local stations, which led, in turn, to diminished quality of programming. The complaint also alleged that the presence of CATV systems inhibited broadcasters' ability to extend coverage and increase advertising revenue through the development of satellite stations.[14]

According to the broadcasters, the issue could be resolved if the FCC were to regulate CATV as a common carrier. It seems, however, that broadcasters were operating under some misconceptions concerning the nature of common carrier regulation because, in addition to its request for more traditional common-carrier controls, the complaint urged the FCC to "determine under what circumstances and for what periods of time CATV systems should be allowed to operate."[15] The exercise of such powers would more closely resemble a system of licensing, something the FCC is not explicitly empowered to do under its Title II common carrier authority.

These legal considerations were not lost on the FCC staff when they issued recommendations to the commissioners. In an August 1957 memorandum, the general counsel, the chief of the Common Carrier Bureau, and the chief of the Broadcast Bureau all recommended that the FCC reject the complaint. Although the staff acknowledged that "upon certain reasonable assumptions" CATV systems might be deemed common carriers in interstate commerce, they concluded that imposing Title II regulation would "serve no reasonable or practical purpose."[16] Common-carrier regulation would have allowed the FCC to regulate CATV consumer rates, but it would not have given the agency authority to restrict the operation of CATV systems. For this reason, common carrier regulation would not provide any real economic protection for broadcasters.[17]

When the FCC closed the matter in the spring of 1958, the request for regulatory proceedings was denied. The FCC's report generally followed the staff recommendations, with one important exception: the commission explicitly rejected the idea that CATV could ever be considered a common carrier. The commission concluded that CATV was not a common carrier in the "ordinary sense" because the CATV system, not the customer, selected the material that was ultimately transmitted.[18] The FCC seems to have based this conclusion on a very narrow definition of a common carrier—that which most closely fit traditional telephone service. In settling on this characterization of common carriers, however, the FCC cited no case law or statutory definitions and ignored the reception-service common carrier category discussed in the 1952 staff memo.

The FCC's disposition of *Frontier Broadcasting* sent the message that the Doerfer commission preferred a narrow reading of statutory authority and thus would approach any claim of regulatory jurisdiction with a skeptical eye. The case of *Frontier Broadcasting*, however, had presented the FCC with a fairly narrow set of legal questions and therefore had not afforded the agency an opportunity to fully evaluate the economic impact of CATV on broadcast television. To gain a broader understanding of the regulatory questions posed by CATV, the FCC opened a special economic impacts inquiry shortly after *Frontier Broadcasting* was completed.

THE ECONOMIC IMPACTS INQUIRY, 1958–1959

Begun in the spring of 1958, the economic impacts inquiry was meant to serve as a broader study of "auxiliary" services, a designation which included CATV, boosters, and translators. Like CATV, both boosters and translators were capable of extending broadcast signals to distant communities; nevertheless, as the inquiry soon revealed, CATV posed the most significant economic threat in the eyes of broadcasters.

Because it lacked CATV expertise, the FCC relied almost exclusively on data obtained from public comments submitted by television broadcasters, CATV systems operators, and the interest groups that represented them. Overall, however, the public comments contained only a modest amount of empirical evidence. Broadcasters presented mainly anecdotal evidence, which consisted of little more than narratives suggesting that some local stations had been harmed by the presence of CATV in their service areas. CATV interests, for their part, offered logical refutations of these claims, but again, little hard evidence.

Perhaps the most notable portions of the public comments consist in the competing rhetorical strategies used by broadcasters and CATV groups. Broadcasters, on the one hand, argued that they held special public interest obligations that needed to be preserved through regulation. For instance, UHF broadcasters—such as the Southern Idaho Broadcasting Company, owner of KLIX-TV of Twin Falls—viewed cable television as a challenge to "the Commission's own system of priorities" in television regulation. As they went on to note, " if KLIX-TV and stations similarly situated fail, the commission's *Sixth Report and Order* will also fail."[19] The National Association of Broadcasters (NAB) stated similarly that CATV was a threat to "the Commission's traditional concern for local service and local programming."[20] The NAB went on to charge that CATV systems were engaged in "unfair competition" because they were essentially rebroadcasting the signals of local stations without consent, an action that violated Section 325(a) of the Communications Act. This was "unfair," of course, because

CATV systems were making free use of the programming for which broadcasters had paid. For the NAB, the solution was to require CATV systems to seek "retransmission consent" from stations before they could carry their signals. [21]

CATV interests, on the other hand, sought to portray their industry as little more than a supplemental service designed to fulfill the most important goal of the *Sixth Report and Order*, namely, the nationwide dissemination of television service. For instance, the Missoula TV Cable Company asserted that broadcasting and CATV were "compatible elements of the broad field of rendering television available to the greatest number of people."[22] The National Community Television Association (NCTA), likewise, asserted that CATV was not rebroadcasting television signals because it was simply a "master antenna" service that was not engaged in broadcasting, per se.[23] Thus, according to the proponents of CATV, the broadcasters' desire to regulate CATV—and not the operation of CATV itself—was truly anticompetitive, serving only to "inhibit the spread of television and restrain free enterprise."[24]

The rhetoric presented by respective broadcasting and CATV interests was repeated in numerous FCC proceedings over the next seven years. Within the context of this inquiry, however, the FCC seemed less interested in broad questions of values than in making sense of the admittedly limited empirical data relating to the economic impact of CATV. Of the ninety-six UHF stations that had gone off the air since 1952, the FCC was able to identify only three instances in which CATV might have had a notable economic impact. In all three cases, however, a CATV system had already been operating in the area when the broadcast stations went on the air. The FCC found it difficult, therefore, to relate changes in the stations' revenue streams to the presence of CATV. Furthermore, the FCC could not easily isolate the impact of CATV from other economic considerations including the size of the television market, the presence of VHF-UHF intermixture, and the availability of UHF television sets.[25] Despite the FCC's ultimate conclusion that CATV likely produced an economic impact of "substantial size," it was not possible to define with any precision the point at which this impact threatened a station's profitability.[26]

Although the FCC did believe that some regulation of CATV was desirable, the commission once again concluded that, under existing statutes, it lacked the jurisdiction to impose any controls. The agency reaffirmed the position it had taken in *Frontier Broadcasting*: CATV was not a common carrier under Title II of the Communications Act. The FCC also rejected the argument of broadcasters that CATV constituted "rebroadcasting" under Section 325(a). Because CATV was delivered over wires, the commission concluded, it could not be considered broadcasting in the traditional sense. Thus, retransmission consent requirements could not be imposed under the law as it stood at the time.[27]

The FCC pledged to ask Congress for the authority both to impose retransmission consent requirements on CATV systems and to require them to carry local television signals upon request from the broadcaster. By supporting retransmission consent and the "must-carry" stipulation, the Doerfer commission sought to address the concerns of broadcasters without imposing undue restraints on free enterprise. Retransmission consent allowed the FCC to guard the property rights of broadcasters without putting in place a complicated system of regulations. CATV systems would seek consent; broadcasters, desiring to extend their service areas, would likely provide it. At the same time, the must-carry requirement helped ensure the economic viability of UHF stations. CATV carriage would extend the reach of local signals while eliminating some of the technical difficulties that made CATV customers less likely to view local stations. No restraints were proposed, however, either to limit the ability of CATV systems to "compete" with broadcasting by importing additional signals or to otherwise control their content.

THE FIGHT FOR LEGISLATION, 1959–1960

In 1959 the FCC's legislative recommendations were introduced in the Senate as S. 1801. These recommendations clearly reflected the Doerfer commission's desire to maximize free enterprise and minimize regulation through a narrow reading of statutory authority. However, when it shifted the CATV issue into the legislative arena, the FCC found itself confronting a challenging political environment where legislation seemed almost doomed to fail. For one thing, Congress knew no more about CATV than the FCC did; the legislature had only begun learning about CATV one year earlier through a set of Senate hearings. Dominated by anecdotal evidence, the hearings painted a complicated picture of the CATV issue, as local broadcasters and constituents expressed a diverse array of views on the matter.[28]

Amid congressional ambivalence, the FCC was unable to advance its own policy preferences because it lacked credibility with Congress. Throughout the 1950s, Congress had become increasingly frustrated with the FCC's inability to forge a regulatory solution to the UHF problem.[29] Then, in the late 1950s, the FCC was criticized by Congress both for its role in a controversial antitrust settlement involving AT&T and for failing to take a strong stand following the quiz show scandals.[30] With the FCC so politicized at the time, any claims the agency made to bureaucratic expertise held little sway with Congress.

Given these political realities, it is not surprising that the FCC's CATV legislation easily ran into a wall of interest group opposition. CATV interests were firm in their opposition to retransmission consent, believing that if broadcasters were granted such explicit property rights over their transmissions, they could essentially put an end to CATV simply by refusing to grant carriage of their signals.[31] Broadcasters, however, argued that retransmission consent alone was insufficient protection; they sought even more regulation. Many western broadcasters, for instance, wanted to augment the original FCC proposal with a requirement that CATV systems be subject to public interest licensing requirements. Believing that some regulation was imminent, the NCTA countered with a softer proposal that would allow the FCC to "certify" new CATV systems but that made no provision for either public interest or retransmission consent requirements.[32]

In an effort to forge a compromise, Senate Commerce Committee chairman John Pastore introduced a bill that incorporated elements of both the broadcaster and CATV proposals. Under this plan, CATV systems would be subject to public interest licensing, but existing systems would be grandfathered through the granting of "automatic" licenses. In addition, CATV systems would be required to carry local broadcast signals, and the FCC would formulate regulations providing for "non-duplication" of local station content. The bill favored CATV inasmuch as it contained neither retransmission consent provisions nor an explicit grant of property rights to broadcasters. Whereas broadcasters quietly supported the bill, however, the NCTA came out in strong opposition to it, citing the must-carry and non-duplication requirements as major concerns. Ultimately, the bill died on the floor of the Senate in the spring of 1960.[33]

Conclusion

As the 1950s drew to a close, federal regulation of CATV seemed to be an unlikely prospect. Rejecting the regulatory activism of the 1940s, John Doerfer and his fellow commissioners preferred a narrow reading of the Communications Act—a reading that limited the scope of FCC regulation. However, given the state of knowledge about CATV at the time, it would be inappropriate to conclude that the Doerfer commission's policies were simply the product of a rigid regulatory philosophy. In the 1950s, CATV was increasing access to television service and promoting content diversity—two of the main goals underlying the *Sixth Report and Order*. Although broadcasters raised concerns that CATV was undermining localism, there was little empirical evidence to support these claims. The Doerfer commission was willing to impose regulations that protected broadcasters' property rights and ensured public access to local signals, but only under an explicit legislative grant of authority. Faced with a complex and conflicted political environment, Congress reached a stalemate, and no such legislation was forthcoming.

Thus, in 1960 three main factors were complicating the process of formulating a CATV policy: (1) limited knowledge concerning the economic impact of CATV on broadcasting, (2) political stalemate among competing interests, and (3) a regulatory philosophy at the FCC that guarded against regulatory activism. During the next five years, the state of knowledge would change only slightly while the political climate would not change at all. However, a shift in the FCC's regulatory philosophy would lead to the first efforts to regulate CATV within just a few years.

Endnotes

1. Don R. LeDuc, *Cable Television and the FCC: A Crisis in Media Control* (Philadelphia: Temple University Press, 1973), 86.
2. *J. E. Belknap*, 18 FCC 642 (1954).
3. US Federal Communications Commission, *Twenty-Fifth Annual Report to Congress* (Washington, DC: GPO, 1959), 53.
4. Lawrence W. Lichty, "The Impact of FRC and FCC Commissioners' Backgrounds on the Regulation of Broadcasting," *Journal of Broadcasting* 6 (1962): 97.
5. Gerald V. Flannery, ed., *Commissioners of the FCC: 1927–1994* (Lanham, MD: University Press of America, 1995), 105–107.
6. Ibid., 105; See also Patrick R. Parsons, *Blue Skies: A History of Cable Television* (Philadelphia: Temple University Press, 2008), 142.
7. Senate Committee on Interstate and Foreign Commerce, *Television Inquiry—Part VI: Hearings before the Committee on Interstate and Foreign Commerce*, 85th Cong., 2nd Sess., 1958, 3507.
8. US Federal Communications Commission, Report and Order, 25 June 1956, Docket 11532 (FCC Docketed Case Files, RG 173, Washington, DC: National Archives).
9. Senate Committee on Commerce, *Appointments to the Regulatory Agencies: The Federal Communications Commission and the Federal Trade Commission*, report prepared by James M. Graham and Victor H. Kramer, 94th Cong., 1st Sess., 1976, Committee Print, 245.
10. Robert E. Lee, *In the Public Interest: The Life of Robert Emmet Lee from the FBI to the FCC* (Lanham, MD: University Press of America, 1996), 165.
11. Kimberly A. Zarkin and Michael J. Zarkin, *The Federal Communications Commission: Front Line in the Culture and Regulation Wars* (Westport, CT: Greenwood Press, 2006), 179.
12. Ibid., 167.
13. *Frontier Broadcasting*, 24 FCC 251 (1958).
14. Ibid., 253.
15. Ibid., 251.

16. FCC Staff, Inter-office Memorandum, 1 August 1957, Reprinted in Senate, *Television Inquiry—Part VI*, 4146.
17. Ibid., 4145.
18. *Frontier Broadcasting*, 254.
19. Southern Idaho Broadcasting, Comments, 7 July 1958, Docket 12443 (FCC Docketed Case Files, RG 173; Washington, DC: National Archives), 9.
20. National Association of Broadcasters, Comments, 7 July 1958, Docket 12443 (FCC Docketed Case Files, RG 173; Washington, DC: National Archives), 4.
21. Ibid., 15–16.
22. Missoula TV Cable Co., Inc., Comments, 26 June 1958, Docket 12443 (FCC Docketed Case Files, RG 173; Washington, DC: National Archives), 3–4.
23. National Community Television Association, Reply Comments, 28 July 1958, Docket 12443 (FCC Docketed Case Files, RG 173; Washington, DC: National Archives), 23.
24. Ibid., 17.
25. *Economic Impacts Inquiry*, Report, 26 FCC 403, 415–418 (1959).
26. Ibid., 422–423.
27. Ibid., 428–430.
28. LeDuc, *Cable Television*, 88–95.
29. Michael J. Zarkin, "Organizational Learning in Novel Policy Situations: Two Cases of United Stated Communications Regulation," *Policy Studies* 29 (2008): 94.
30. Senate Commerce Committee, *Appointments*, 93; Susan L. Brinson, "Epilogue to the Quiz Show Scandal: A Case Study of the FCC and Corporate Favoritism," *Journal of Broadcasting and Electronic Media* 47 (2003): 279–281.
31. LeDuc, *Cable Television*, 99–100.
32. Parsons, *Blue Skies*, 161.
33. Ibid., 164–169.

Chapter 5

Regulatory Experimentation, 1961–1969

Speaking before Congress in 1965, Commissioner Lee Loevinger candidly stated his belief that the FCC's recent about-face concerning CATV regulation "was induced not by any new investigations or any new review of facts, but simply by a change of personnel."[1] Loevinger's statement is a slight exaggeration, at best. Between 1961 and 1963, a group of liberal Democrats appointed by President Kennedy ascended to the commission and brought with them a strong belief in localism and a zeal for regulatory activism. Concerned that CATV might undermine local broadcasting, these "New Frontier" commissioners took preemptive action, imposing regulation without substantial evidence of economic impacts.

Yet, within a short time this activist FCC realized that its initial approach to regulation was both administratively unworkable and practically inapplicable to a changing industry. Whereas in the 1950s the average CATV system had been capable of carrying only three channels, by the end of the 1960s, twenty-channel systems were on the horizon—a change which created the potential for CATV to become the delivery mechanism for numerous new types of television content. As the 1960s ended, the FCC was struggling to find a balance between protecting broadcasting and promoting the development of what was becoming known as the "cable television" industry.

New Frontier Commission

The early 1960s was a time of renewed regulatory activism at the FCC. During these years, the agency was dominated by liberal Democrats who believed that localism was the predominant concern in broadcasting. Competition in broadcasting was desirable, but only if it consisted of local broadcaster competing against local broadcaster for both viewers and advertising revenue. Because CATV systems imported distant signals, they placed local broadcasters in competition with stations from outside the local area and, thus, were a disruptive force in the preservation of established regulatory principles. Real or imagined, the fear that CATV might undermine localism was the motivating factor behind the decision to regulate.

The first steps toward the implementation of this regulatory philosophy were taken in 1961 when Newton Minow was confirmed as FCC chairman. A young lawyer with ties to the Kennedy family, Minow came to the FCC with virtually no formal experience in the communications field.[2] What he lacked in experience, however, he made up for in vision. In his Senate confirmation hearing,

Minow declared that "the Commission has a role in elevating and encouraging better programs, and I am determined to do something about it."[3] For Minow, the key to improved programming was increased competition among local broadcasters, and he believed this aim could be reached by encouraging the development of UHF.[4] Although Minow left the FCC after only two years, his basic vision for television was carried forward under the leadership of his successor, E. William Henry.[5]

Although both Minow and Henry were strong believers in localism, neither man arrived at the FCC with a regulatory agenda for CATV. That void was filled when Kenneth Cox joined the commission in early 1963. Cox had served as a staff member for the Senate Commerce Committee in the 1950s, and during that time he had formed the opinion that CATV should be regulated. As a committee staff member, Cox prepared a report in which he concluded that localism was the most important priority articulated in the FCC's *Sixth Report and Order*. Because CATV had the potential to draw revenue away from local stations, it presented a threat to established regulatory goals. Furthermore, the smallest communities could support neither a local station nor a CATV system, and frequently these communities received their only television signal from a nearby city. Cox concluded that if local stations in medium-sized communities went off the air because of CATV, smaller communities would lose access to television service altogether.[6] Cox believed both that the FCC possessed jurisdiction over CATV under the Communications Act and that the agency should use its authority to forestall these problems.[7]

Cox was elevated to the commission after serving a brief stint as the FCC's Broadcast Bureau chief. Upon his nomination to the commission, Cox was characterized by a prominent trade publication as "a liberal who advocates broad federal regulation of radio and TV stations" and someone who "has definite ideas, for which he will fight hard."[8] This description accurately describes Cox's time as Broadcast Bureau chief, where he was a fierce advocate for localism.[9] With respect to CATV, his convictions were equally firm. Years later, a senior staff member who had served at the FCC during this time noted that Cox "came … knowing full well what cable was and he came with an agenda. And his agenda … was: 'I want that local station carried and I want its programming not duplicated.' "[10] Not surprisingly, local must-carry and non-duplication requirements became the backbone of the 1965–1966 CATV rules.

Although Cox managed to secure the support of a bipartisan majority of his colleagues, he faced strong opposition to his CATV agenda from fellow Democrat Lee Loevinger. In testimony before Congress, Loevinger distinguished himself from his colleagues by stating that he had "far more confidence … in the ability of the economy, in the ability of the public, to secure the results it wants regardless of what government does."[11] He strongly objected to the 1965–1966 CATV rules, believing them to be little more than an effort to protect the economic interests of broadcasters.[12] Loevinger believed that the appropriate role for the government was to promote technological development and to avoid interfering with consumer choice.[13]

Carter Mountain: The First Step

The FCC's changed regulatory attitude toward CATV first became apparent in its 1962 ruling in the *Carter Mountain* case, which involved little more than a routine request for a license to provide common carrier microwave relay service to CATV systems. Since the *Belknap* decision in 1954, the FCC had consistently granted such licenses to applicants who could demonstrate that they were bona fide common carriers. This meant only that the carrier was technically capable of providing service and was willing to offer it to all customers at reasonable charges and without discrimination. Economic impact on television service was not a consideration in microwave-licensing decisions of this kind.[14]

In 1959, however, the situation changed when KWRB-TV decided to protest the FCC's approval of a microwave license application submitted by the Carter Mountain Transmission Corporation of northern Wyoming. In asking the FCC to reconsider the license, the station's primary argument was grounded in the rationale of economic impact. KWRB's chances of success seemed slim, however, given that the FCC had recently completed its economic impacts inquiry and had so far been unwilling to use existing statutory authority to impose regulation on CATV. Nevertheless, KWRB sought to portray the Carter Mountain application as a special case, arguing that the economic impact from distant-signal importation would likely force KWRB to go off the air, eliminating the only television signal available to thousands of rural residents who did not receive CATV service.[15]

Initially, the FCC staff supported a denial of KWRB's protest. In a brief signed by the chiefs of both the Common Carrier and Broadcast Bureaus, Carter Mountain was declared a bona fide common carrier under existing FCC precedents. However, by the time the application was brought forward for oral argument late in 1961, an important staff change had caused the Broadcast Bureau to break ranks. Kenneth Cox was now the bureau chief, and he urged the commission to consider the economic impact in this particular case. Speaking before the full commission, staff member Robert Rawson stated the Broadcast Bureau's position:

> It is our belief that the Commission has never said … that it will not take into consideration the impact that the authorization of the microwave relay facility which would carry a CATV system … would have upon the ability of

> a television station to survive ... If the Commission is to preserve the integrity of the priorities that are laid down in television ... it must consider this impact.[16]

Rawson went on to state that the FCC's conclusions in the 1959 economic impacts inquiry were simply policy statements and not binding precedent. Thus, the commission could and should consider amending them or abandoning them altogether.[17]

Ultimately, the FCC accepted the Broadcast Bureau's recommendations and denied the Carter Mountain license, noting that "the well-being of existing television facilities is an aspect of [the] public interest."[18] The commission invited reapplication only if Carter Mountain guaranteed that CATV systems employing their services would carry local signals and would not duplicate local station programming.[19]

Protecting Small Communities: The 1965 Rules

The *Carter Mountain* case seemed to validate Kenneth Cox's conviction that CATV had the potential to undermine localism and destroy television service in rural areas. However, imposing must-carry and non-duplication requirements on a few CATV systems through a single microwave-licensing proceeding would do little to protect television service nationally. Recognizing the need for generally applicable regulations, the FCC requested public comment on a proposal to extend local-carriage and non-duplication requirements to all CATV systems employing microwave relay service.[20]

Broadcast and CATV groups responded to the request with rhetorical positions quite similar to those advanced during the economic impacts inquiry several years earlier. This time, however, broadcasters also came armed with policy analysis aimed at demonstrating harm to local stations in smaller communities.

The NAB commissioned Dr. Franklin Fisher, an economics professor at the Massachusetts Institute of Technology (MIT), to do a thorough study of advertising revenue in 136 television markets with one or two local stations. Fisher concluded that a "substantial" decrease in revenue was caused by CATV, particularly in single-station markets where the local signal was not being carried by the CATV system. The revenue losses decreased when there was either competition from a second local station or signal carriage by the local CATV system, or both.[21] Fisher also noted substantial revenue losses when duplication of network television programming occurred.[22]

In spite of the NAB's contention that Fisher's study presented "solid, documented, factual information,"[23] CATV interests quickly took issue with his findings. The NCTA hired Dr. Herbert Arkin of City College in New York to provide an evaluation of the Fisher study. Arkin characterized Fisher's findings as "at best of dubious validity and at worst a complete misstatement."[24] Although Arkin conceded that revenue losses were severe when local stations were not carried by the CATV system, he pointed out that there were relatively few instances of non-carriage in Fisher's sample.[25] Using its own sample of 723 CATV systems from across the country, the NCTA found only one single-station market in which CATV was not carrying the local station. But in nearly all non-carriage situations, the CATV system provided special equipment that allowed customers to easily switch over to the local stations.[26] Finally, Arkin and the NCTA found it generally difficult to replicate Fisher's conclusions regarding station revenue losses.[27]

Lacking the expertise to critically assess the NAB and NCTA studies, the FCC appointed economist Martin H. Seiden to conduct an independent analysis. Whereas the NAB and NCTA studies offer sharply contrasting views of CATV's economic impact, the Seiden study paints a more complex picture. On the one hand, Seiden found little evidence that CATV was having an economic impact on broadcasters. From among the eighty-six stations which experienced declining revenues between 1961 and 1963, Seiden found only four with substantial CATV penetration in their service areas. At the same time, however, Seiden allowed for the possibility that the economic impact could become significant later on as the presence of CATV became more apparent to local advertisers.[28] Furthermore, although Seiden concluded that "economic protection alone cannot be the basis of restricting CATV," he allowed for the possibility that must-carry and non-duplication requirements might be used as an interim policy.[29]

For the FCC, these studies did little to clarify the present or future economic impact of CATV on small-market television stations. In the absence of good information, the commission chose to learn by doing; it opted to impose a limited set of regulations as a first step. The 1965 CATV rules required all microwave-fed CATV systems to carry and avoid duplicating the signals of all stations within their immediate service areas.[30] Though these rules applied only to a relatively small portion of all CATV systems, the FCC viewed them as a first step designed to help the agency "build up a body of experience … before considering the extension of the rules to the CATV industry as a whole."[31]

Protecting Large Communities: The 1966 Rules

Before the 1965 rules had been finalized, broadcasters were already pressuring the FCC to design a broader set of regulations that would be applicable to all CATV systems. One of the most urgent calls for action came from the American Broadcasting Company (ABC), which claimed that CATV at this point threatened television stations in urban areas. Because many CATV operations were by that time capable of carrying twelve channels or more, ABC speculated that these systems could import enough distant signals to undermine the profitability of all three network affiliates in major population centers. Over time, this increased profitability would allow CATV to produce its own programming and so replace broadcast television altogether.[32]

Certainly, there were political motivations behind ABC's decision to advance this argument. In the early 1960s, ABC was a struggling television network that needed to acquire more local affiliates in order to survive.[33] This could only happen if UHF flourished, particularly in urban areas where local stations could exist profitably. At the same time, CATV penetration into urban markets threatened to undermine FCC policy goals relating to the All-Channel Television Receiver Act of 1962. The act itself was little more than a single sentence authorizing the FCC to require all television sets made in or imported to the United States to be UHF-ready. Buried deep in the legislative history of the law, however, was the hope Congress expressed that UHF development would make possible the existence of four competing commercial television stations in every large population center.[34] Whether or not Congress actually intended this legislation to be a formal statement of policy, the FCC used it as a justification for following up on ABC's petition with a formal rulemaking proceeding.[35]

Of fundamental importance in the proceeding was the lingering question of whether the FCC possessed statutory jurisdiction to regulate the entire CATV industry. The controls imposed on microwave-fed CATV systems were narrowly justified as regulation of the "end use" of a common carrier facility under Title II of the Communications Act. Thus, regulation of non-microwave CATV systems needed to be justified by invoking some additional grant of statutory authority. In a publicly circulated memorandum, the FCC's general counsel argued that CATV systems were "instrumentalities" of radio communication under the terms of the Communications Act. If this was true, CATV could be regulated when necessary in order to support the goals contained in Title III—including the mandate to ensure a "fair … and equitable" division of broadcast services across the states and communities.[36] In essence, if CATV systems threatened to put broadcasters out of business, they were interfering with the "fair … and equitable" distribution of broadcast services and could therefore be regulated under existing authority. Ultimately, a majority of the commissioners found this rather expansive reading of the Communications Act compelling.

Having thus established legal jurisdiction, the commission seemed satisfied to once again proceed in a precautionary manner, imposing regulation without substantial evidence of the economic impact of CATV on broadcasters. In fact, the commission specifically declined to solicit any further studies of economic impacts, noting that

> studies of this nature … are of limited value since they cannot measure some of the most important factors we are bound to consider. These include the cumulative future effect of greater penetration by CATV systems franchised or applied for but not yet in operation, the degree of success to be achieved by CATV systems in big cities or other well-served areas, and the effect of the burgeoning CATV activity—if left unregulated—on the decisions of the potential [broadcast] applicants and existing licensees as to whether to inaugurate or improve service.[37]

Studies of economic impact were not helpful, the FCC asserted, because the impacts the commission sought to regulate were mostly hypothetical. The "burgeoning CATV activity" that led the

FCC to regulate in 1966 consisted mostly of pending applications for franchises in major cities. At the time, San Diego was the only major city with significant CATV penetration, and no evidence existed to demonstrate any economic impact in that area.[38]

The core features of the 1966 rules were the same as those that had been imposed on microwave systems a year earlier. Any CATV system located within a station's broader reception area was required, upon request, to carry that station's signal and to avoid duplicating its content.[39] To further protect stations in urban markets, the commission also prohibited CATV systems from importing distant signals into the nation's largest one hundred television markets.[40] Recognizing the difficulties posed by a one-size-fits-all approach for all CATV systems, the FCC provided opportunities for CATV operators to seek waivers for both the must-carry and distant-signal importation requirements.

Although a majority of the commissioners voted for the 1966 rules, none were truly enthusiastic in their support. In congressional hearings later that year, both Commissioners Henry and Cox expressed doubt that the regulations could provide a long-term solution to the CATV problem.[41] Commissioner Loevinger, a dissenter in the 1966 decision, was more stinging in his criticisms. He characterized the commission's claim to jurisdiction over CATV as an "intellectual bootstrapping operation"[42] undertaken by commissioners with "strong and fixed views regarding the subordinate place of CATVs in the mass communications system."[43] Loevinger warned further that the non-duplication requirements could lead CATV systems to fill in the "holes" by developing their own programming, something the commission had explicitly sought to avoid in crafting the rules.[44]

Thus, almost as soon as the 1966 CATV rules had been finalized, the FCC was ambivalent about the regulations' future prospects.

THE ROAD TO REASSESSMENT, 1967–1968

Over the next two years, the FCC's ambivalence evolved into a firm conviction that the 1966 rules should be reassessed. This conviction was shaped by three factors, the first of which was a series of court decisions that solidified the FCC's role as the principal actor in CATV regulation. In the case of *US v. Southwestern Cable Co.*, the US Supreme Court affirmed the FCC's statutory jurisdiction over CATV.[45] Drawing on earlier precedents, the Court reiterated that the Communications Act was intended to be a "unified and comprehensive regulatory system" and as such should be interpreted broadly by the agency.[46] The Court also accepted economic harm to broadcasters as a cause for action, noting that the FCC could regulate for any reason that was "reasonably ancillary" to its Title III responsibilities.[47]

In a second case, *Fortnightly Corp. v. United Artists Television*, the Court ruled that federal copyright law did not require that CATV systems request permission to use a broadcaster's signal.[48] Had the ruling gone the other way, broadcasters could have essentially placed a stranglehold on CATV by refusing permission. In that situation, either CATV would have stagnated, or Congress would have altered the law to make it more favorable to the new medium. By maintaining the status quo, however, the Court guaranteed that CATV would continue to grow unabated in the short term. With Congress perpetually stalemated over legislation, the question of whether and how to regulate CATV was now placed squarely on the FCC's shoulders.

A second factor driving reassessment of the 1965–1966 rules was the FCC's experience with implementation. In order to more efficiently handle administrative matters relating to CATV, the FCC set up a special implementation task force in 1966. The CATV Task Force, as it came to be known, consisted of nine lawyers drawn from various staff offices within the agency. By October of that year, the task force faced a backlog of several hundred petitions from CATV systems for must-carry and distant-signal importation waivers.[49] The overwhelming workload led the first chief of the CATV Task Force to characterize his office as a "small, beleaguered staff operating out of a discouraging organizational set-up."[50]

The third—and perhaps most important—factor that led the commission to reassess existing regulations was a series of policy documents presenting a new vision for CATV or, as it was now called, "cable television." One particularly important set of publications emerged from a RAND study highlighting the potential for cable to remake the United States into a series of "wired cities." According to these studies, twenty-channel wire cable systems would someday connect every home in a city, serving more or less as common carriers. Channel capacity would be leased to traditional networks and pay-television services, and space would also be reserved for public interest purposes, such as education and political communication. The RAND publications presented this system as superior to traditional broadcast television in terms of cost, quality of service, and diversity of programming.[51]

A second, similar document was a government report published by the President's Task Force on Communications Policy in 1968. Although it reaffirms the traditional concern for localism, the report also places special emphasis on the potential for television to serve as a medium for education and self-expression in urban communities. By virtue of their large channel capacities, cable systems could play a unique role in bringing more diverse entertainment, political, and instructional content into inner-city areas. The report is cautious, however, noting that sound public policy would be needed to make these potential benefits reality; unfettered cable development might result in a two-tiered system in which the most desirable content is available only to those possessing the means to pay for it.[52]

Together, these reports changed the conversation about cable television in important ways. The new medium held the potential not only to bring subscribers diverse programming through distant signals but also to develop new content that served the public interest. Government policy, therefore, needed to nurture the growth of cable and the development of innovative content. At the same time, however, this new vision for cable fit nicely with traditional concerns for localism and other public interest considerations. Not surprisingly, these ideas resonated with several members of the FCC, including Kenneth Cox and relative newcomer Nicholas Johnson—an LBJ appointee with a strong belief in television as a public service enterprise.[53] As 1968 progressed, the new ideas found their way into commission debates concerning the future of cable television regulation.

Policy Development, 1968

The first indication of a change in thinking came in mid-1968, when the FCC decided the *Midwest Television* case.[54] The matter started out as little more than a request from a San Diego cable provider for a partial waiver of the must-carry requirements. During the course of the proceedings, however, the commission discovered that the San Diego CATV system was originating its own programming, a prospect that had alarmed the commission just two years earlier. By this time, however, the FCC seemed to approve of program origination, noting that

> CATV program origination offers promise as a means for increasing the number of local outlets for community self-expression and for augmenting the public's choice of programs and types of service … Whereas television

> broadcast stations are usually located in or near a central community and were not intended to serve a much broader area … almost every community of an appreciable size could have its own CATV system and thus its own local outlet … The CATV operator has more flexibility to present programming of minority interest on some channels. And, finally, CATV program origination does not entail the question of "unfair competition" posed by CATV importation of broadcast signals, or any disparate situation with respect to copyright liability, and would be less likely to duplicate the programs of local broadcast stations.[55]

In essence, the commission was now of the opinion that CATV program origination offered the best of both worlds: an increase in the number of outlets for local self-expression, particularly in smaller communities, and no economic threat to broadcasters.

Nevertheless, the *Midwest Television* order was more a statement of preference than a broad reformulation of policy. Formal reconsideration of the 1966 regulatory scheme was bogged down amid disagreement within the FCC regarding the proper bureaucratic niche for cable. Some within the agency favored elevating the CATV Task Force to the status of a permanent bureau and then treating cable as a distinct communications medium. Others contended that cable should be regulated through the Broadcast Bureau, where it would be treated as a supplement to traditional television service. Still others believed that cable was evolving into a wire-based video common carrier that would be best regulated through the Common Carrier Bureau.[56] Faced with this disagreement, the commission asked senior staff from different bureaucratic niches to prepare position papers outlining new options for cable regulation.

The first paper was prepared by Sol Schildhause, chief of the CATV Task Force. Prior to heading the task force, Schildhause had been a respected lawyer and hearing examiner within the agency for more than fifteen years. Although he came to the task force with little knowledge of cable, he soon became an advocate for its development. Believing that cable was the future of television, Schildhause sought to establish a more permanent existence for his task force, which at the time was little more than an ad hoc group of bureaucrats assembled to perform routine implementation tasks.[57] The position paper was an important opportunity for Schildhause to both establish his task force's credibility with the commissioners and make the case that the task force could serve as a center for cable policy development.

In his position paper, Schildhause highlighted the need for increased expertise, noting that "despite its experience with existing regulations, the commission lacks sufficient information to conduct long range planning in the CATV field."[58] For Schildhause, the solution was to increase the research and analysis capabilities of the task force.[59] However, this admission of inadequate expertise did nothing to lend credibility to the recommendations that followed, which consisted of a policy designed to promote cable growth. Schildhause suggested a short-term strategy of restricting application of the 1966 distant-signal importation limitations to the largest fifty television markets only. This would reduce the administrative burden on the task force while increasing television content diversity in medium-sized markets.[60] Schildhause also suggested a long-term strategy of promoting cable growth and development in geographical areas where the technology was most needed.[61]

Schildhause's ideas stood in sharp contrast to those of Henry Geller, then the FCC's general counsel and the principal author of a second position paper on cable policy. A career staff member at the FCC, Geller had been promoted to general counsel in the early 1960s; he was one of the principal architects of the

1965–1966 regulations.[62] Like many at the FCC during this time, Geller strongly believed that television was a public service endeavor and that cable could only be valuable to the extent that it, too, served the public interest.

In a position paper coauthored with the chief of the Broadcast Bureau, Geller advocated the absolute prohibition of distant-signal importation in the country's largest one hundred markets. This approach would eliminate the existing administrative burden while providing continued regulatory protection for developing UHF stations in large markets.[63] Under this proposal, the continued existence of cable providers in large markets would depend on their ability to engage in program origination and so serve as an outlet for local expression. To insulate UHF stations from direct competition with cable, Geller proposed a ban on commercial advertising in original cable programming. To support themselves, cable providers would essentially need to develop a two-tiered rate system in which they charged more for certain types of original programming.[64]

Geller's recommendations reflected the best-of-all-worlds scenario envisioned by the FCC in the *Midwest Television* order. Program origination offered a way for cable to grow and serve the public interest without causing economic harm to broadcast television. In essence, the commission had found a way of understanding cable that allowed them to embrace the social and technological potential of the new medium without abandoning existing policy goals or ideas. Possessed of this newfound sense of clarity, the commission was ready to propose new cable regulations late in 1968.

The 1968–1969 Proposed Rules

The 1968 proposed rules integrated the basic themes outlined in the *Midwest Television* order as well as recommendations contained in both the Geller memo and the external policy documents. The FCC proposed to condition cable carriage of broadcast signals on a requirement that systems with over 3,500 subscribers engage in program origination. The commission further proposed that cable programming should be required to observe the fairness doctrine and other public interest requirements that already applied to broadcast licensees. Larger cable systems would also be required to provide local access channels for educational, governmental, and public use.[65]

With respect to distant-signal importation into large markets, the FCC proposed to eliminate all hearings and waivers and to require cable operators across the board to obtain retransmission consent. Initially, this move appeared to be a significant departure from the absolute ban on signal importation proposed by Geller, because cable operators who managed to obtain retransmission consent would apparently be allowed to import unlimited distant signals. However, a clarification issued by the FCC in early 1969 revealed that retransmission consent actually posed a significant obstacle to cable's use of broadcast signals. Cable operators would have to obtain program-by-program consent from all parties holding a property interest in the programming—parties which might include not only a station but also a network, a distributor, and a producer.[66] In an unusual move, the FCC also announced that it would only entertain pending distant-signal importation waivers if broadcast stations agreed to retransmission consent, effectively making the proposed rule immediately binding on an interim basis.

By imposing retransmission consent rather than a complete ban on signal importation, the commission hoped to encourage broadcasters, cable operators, and program suppliers to negotiate a copyright agreement, filling the legal void left by the US Supreme Court.[67] If these three industries could agree among themselves to grant general retransmission consent in exchange for copyright payments, cable could then import signals into urban areas without being accused of unfair competition. If the industries could not agree, distant-signal importation would remain frozen in the largest one hundred markets for the foreseeable future.

Conclusion

Between 1961 and 1969, the evolution of CATV into cable television was accompanied by an evolution in the FCC's understanding of the medium. From 1961 to 1966, CATV was capable only of enhancing broadcast signals and supplementing them through distant-signal importation. Fearing that broadcasters might suffer adverse economic impacts from CATV, the FCC acted with precaution by choosing to regulate even in the absence of clear empirical evidence. Between 1967 and 1969, practical experience led the FCC to conclude that the existing regulations were unworkable, while new ideas pointed the way to alternative solutions. Cable television was by this time a medium capable of delivering new content to urban communities and serving the public interest in the process. Public policy needed to promote the development of this new medium, rather than simply curtail cable's impact on broadcasters.

Even though this evolution in thinking helped the FCC move forward on cable television issues, the shift did not mark a dramatic departure from past ideas. The new perspectives—presented by the President's Task Force report and the RAND study—resonated with the FCC precisely because they reinforced traditional conceptions of the public interest in television. Cable television was viewed as a medium capable of delivering a broad array of entertainment, educational, and public affairs content and of serving as an outlet for local expression—a set of aspirations quite similar to those delineated for broadcasting in the 1940s. Furthermore, the 1968–1969 regulatory proposal did not depart significantly from existing practices in communications regulation. The program origination and public interest requirements treated cable like a broadcaster, while the proposed public access channels made portions of a cable system into a quasi-common carrier for content developed at the local level. In essence, the commission was trying to find an amorphous niche for cable somewhere between broadcasters and common carriers.

Yet, even while the FCC was trying to promote cable's potential, concern over the economic impact on broadcasters continued to dominate policy deliberations. Retransmission consent requirements were restrictive and made it virtually impossible for cable providers to build up a subscriber and revenue base through distant-signal importation. It was unclear whether, without a substantial revenue base, cable systems could undertake the program origination that was supposed to be their lifeblood in this new era.

Furthermore, the FCC continued to develop policy in an environment characterized by limited information. In 1968 the FCC knew no more about the extent of economic impacts than it had known two years earlier. Nor, for that matter, did the agency at that point really understand the economic or technical considerations associated with program origination in cable television. The agency's analytical capacities remained weak, and policy recommendations were developed by a diverse group of senior staff who often found themselves at loggerheads.

Thus, as the FCC moved forward with its new vision for cable television, its prospects for immediate success seemed uncertain.

Endnotes

1. House Committee on Interstate and Foreign Commerce, *Regulation of Community Antenna Television: Hearings before the Subcommittee on Communications and Power*, 89th Cong., 1st Sess., 1965, 41.
2. James L. Baugham, *Television's Guardians: The FCC and the Politics of Programming, 1958–1967* (Knoxville: University of Tennessee Press, 1985), 57.
3. Senate Interstate and Foreign Commerce Committee, *Nomination of Newton Minow to be a Member of the Federal Communications Commission*, 87th Cong., 1st Sess., 1961, 7.
4. Baugham, *Television's Guardians*, 87, 92.
5. Gerald V. Flannery, ed., *Commissioners of the FCC, 1927–1994* (Lanham, MD: University Press of America, 1995), 130.
6. Senate Committee on Interstate and Foreign Commerce, *The Problem of Television Service for Smaller Communities*, report prepared by Kenneth A. Cox, 85th Cong., 2nd Sess., 1958, Committee Print, 4–5, 33–36.
7. See his statement in House, *Regulation of Community Antenna*, 56.
8. *Broadcasting*, "Cox Appointment gives FCC a Club Flush," 63, no. 25 (1962): 58.
9. Baugham, *Television's Guardians*, 106.
10. Henry Geller, Oral History (Denver CO: The Cable Center).
11. House, *Regulation of Community Antenna*, 66.
12. See Loevinger's dissenting opinion in US Federal Communications Commission, Memorandum Opinion and Order, 20 January 1967, Dockets 14895, 15233, and 15971 (FCC Docketed Case Files, RG 173; Washington, DC: National Archives).
13. *Broadcasting*, "Forget Your CATV Fears, Loevinger Says," 68, no. 24 (1965): 74.
14. *Intermountain Microwave*, 24 FCC 54 (1958).
15. KWRB-TV, Protest and Petition for Reconsideration, 27 May 1959, Docket 12391 (FCC Docketed Case Files, RG 173; Washington, DC: National Archives), 3–7.

16. US Federal Communications Commission, Statement of the Broadcast Bureau, Oral Argument, 14 December 1961, Docket 12391 (FCC Docketed Case Files, RG 173; Washington, DC: National Archives), 383–384.
17. Ibid., 386.
18. *Carter Mountain* 32 FCC 459, 461 (1962).
19. Ibid., 465.
20. US Federal Communications Commission, Notice of Proposed Rulemaking, 14 December 1962, Docket 14895 (FCC Docketed Case Files, RG 173; Washington, DC: National Archives).
21. Franklin M. Fischer, *The Impact of CATV Competition on Local Television Service*, 26 October 1964, Dockets 14895 and 15233 (FCC Docketed Case Files, RG 173; Washington, DC: National Archives), 1–2.
22. Ibid., 83.
23. National Association of Broadcasters, Reply Comments, 26 October 1964, Dockets 14895 and 15233 (FCC Docketed Case Files, RG 173; Washington, DC: National Archives), 13.
24. National Community Television Association, Additional Reply Comments in Opposition to Proposed Rulemaking, 14 December 1964, Dockets 14895 and 15233 (FCC Docketed Case Files, RG 173; Washington, DC: National Archives), 6.
25. Ibid.
26. Ibid., 10–11.
27. Ibid., 6–8.
28. Martin H. Seiden, *An Economic Analysis of Community Antenna Television Systems and the Television Broadcasting Industry* (Washington, DC: GPO, 1965), 74.
29. Ibid., 73, 89.
30. *Report and Order in Dockets 14895 and 15233*, 38 FCC 683 (1965).
31. Ibid., 687.
32. American Broadcasting Company, Petition for Rulemaking, 16 October 1964, Docket 15971 (FCC Docketed Case Files, RG 173; Washington, DC: National Archives), 4–5.
33. Indeed, by the early 1960s, this was a longstanding argument in FCC proceedings. See Baugham, *Television's Guardians*, 90.

34. House Committee on Interstate and Foreign Commerce, *All Channel Television Receivers*, 87th Cong., 2nd Sess., 1962, H. Rep. 1559, 3.
35. *Notice of Inquiry and Notice of Proposed Rulemaking in Docket 15971*, 30 FR 6078, 6079 (1965).
36. The memorandum is reprinted in *Second Report and Order* 2 FCC 2d 793–797 (1966).
37. *Second Report and Order* 2 FCC 2d 725, 744 (1966).
38. Ibid., 741, 743. One broadcaster in the area alleged that audience share for local stations had been reduced by 25 percent or more because of CATV, but it was not clear that this actually translated into reduced revenue for the station.
39. Ibid., 746.
40. Ibid., 782.
41. Their positions are summarized in Don R. LeDuc, *Cable Television and the FCC: A Crisis in Media Control* (Philadelphia: Temple University Press, 1973), 155.
42. US Federal Communications Commission, Memorandum Opinion and Order, 20 January 1967, Dockets 14895, 15233, and 15971 (FCC Docketed Case Files, RG 173; Washington, DC: National Archives), 4.
43. US Federal Communications Commission, Notice of Inquiry and Notice of Proposed Rulemaking, 22 April 1965, Docket 15971 (FCC Docketed Case Files, RG 173; Washington, DC: National Archives), 5.
44. Ibid., 7.
45. *US v. Southwestern Cable Co.*, 392 US 157 (1968).
46. Here, the court cited *FCC v. Pottsville Broadcasting Co.*, 309 US 134, 137 (1940).
47. *Southwestern Cable*, 178.
48. *Fortnightly Corp. v. United Artists Television*, 392 US 390 (1968).
49. *Broadcasting*, "Mass Action Set for CATV," 71, no. 18 (1966): 52.
50. US Federal Communications Commission, CATV Task Force Chief, Inter-office Memorandum, 30 July 1968 (Papers of commissioner Rosel Hyde, MSS 142; Provo, UT: Brigham Young University Special Collections), 3.
51. For a summary of the study's findings by its authors, see Harold J. Barnett and Edward Greenberg, "On the Economics of Wired

City Television," *American Economics Review* 58 (1968): 503–508.

52. President's Task Force on Communications Policy, *Final Report* (Washington, DC: GPO, 1968).
53. Patrick R. Parsons, *Blue Skies: A History of Cable Television* (Philadelphia: Temple University Press, 2008), 252–254.
54. *Midwest Television*, 13 FCC 2d 478 (1968).
55. Ibid., 505–506.
56. *Broadcasting*, "FCC Starts Work on Basic Cable Policy," 75, no. 7 (1968): 58.
57. Sol Schildhause, Oral History (Denver, CO: The Cable Center).
58. CATV Task Force, Inter-office Memorandum, 2.
59. Ibid., 4.
60. Ibid., 1–2.
61. Ibid., 21–24.
62. Geller, Oral History.
63. US Federal Communications Commission, General Counsel and Chief, Broadcast Bureau, Inter-office Memorandum, 30 July 1968 (Papers of commissioner Rosel Hyde, MSS 145; Provo, UT: Brigham Young University Special Collections), 5–8.
64. Ibid., 13–14.
65. *Notice of Proposed Rulemaking in Docket 18397*, 15 FCC 2d 417, 422–427 (1968).
66. Parsons, *Blue Skies*, 255.
67. Sol Schildhause, "CATV—From the Regulatory Point of View," *Public Utilities Fortnightly* 25 September 1969, 79.

Chapter 6

Negotiating Comprehensive Regulation, 1969–1972

By 1972, the FCC had succeeded in developing a comprehensive new scheme of regulations for cable television. These rules were rooted in the FCC's vision of cable as a hybrid of both the broadcasting and common carrier models, a novel medium capable of providing innovative content and serving as an important new outlet for community expression. The rules reflect the FCC's longstanding interest in balancing the promotion of cable with the economic protection of broadcasters; nevertheless, the details contained in the final product bear only a partial resemblance to the 1968 proposal. The 1972 rules allowed some distant-signal importation into the largest urban areas, turning six years of FCC policy on its head. At the same time, the regulations granted broadcasters some new protections designed to prevent content duplication as cable entered the program origination business. Mandatory cable program origination moved forward, and cable systems were required to maintain special channels for educational, governmental, and public access use. All parties agreed, moreover, to support copyright legislation as a way to permanently resolve property rights controversies arising from cable's use of broadcast signals.

The new rules were not the result of any newfound cable television experience or expertise on the part of the FCC. Rather, they represented a negotiated compromise among broadcasters, cable providers, and copyright owners. Although these interests had been negotiating independently throughout 1969, compromise was facilitated by the leadership of FCC chairman Dean Burch, who found new ways to bring the groups together and initiate constructive conversation. In many respects, Burch's leadership and the resulting 1972 rules were indicative of the FCC's taking a more pragmatic approach to cable television issues than it had in the past.

Interest Group Deliberations, 1969

With restrictive retransmission consent requirements hanging over their heads, cable providers sought to negotiate an agreement with broadcasters and program suppliers that would allow cable into urban areas while forcing them to respect the property rights of the other interested parties. In mid-1969, NAB and NCTA staff prepared a proposed pact designed to do just that. Cable operators were asked to accept some copyright liability for their use of broadcast programming, to carry local stations, and to forego the right to interconnect their systems. In return, cable providers would be allowed to import enough distant signals to provide their service area with three network affiliates and three independent stations. In addition, cable would be allowed to originate one channel of entertainment programming with commercial advertising.[1]

Although the NAB and NCTA had assumed that compromise would be necessary, both groups were initially optimistic about the prospect of an eventual agreement; the reactions of other key industry players, however, did not bode well. The Association of Maximum Service Telecasters (AMST), a broadcast industry group, objected to the signal importation provisions in the proposed pact, contending that they would impair the development of UHF. ABC took matters a step further when it charged that the agreement would harm the profitability of all broadcast television and undermine program diversity and local self-expression throughout the country. Program suppliers were no more enthusiastic about the settlement; they believed that the copyright provisions were too favorable to cable. Whereas cable interests were generally more positive regarding the proposal, some operators did object to the blanket restriction on interconnection, which limited their ability to televise programming to a broad national audience. Amid a sea of interest group conflict, the NAB rejected the proposal, and by September, negotiations were virtually dead.[2]

With the prospect of an agreement stalled, cable providers gradually began the march toward becoming program originators. Suppliers were marketing special new cable programming packages designed to appeal to customers with diverse tastes. The NCTA also proposed a plan to develop a network of six non-entertainment channels and deliver it to cable systems around the country via satellite. The channels would provide space for public broadcasting, for a twenty-four-hour weather channel, and for live telecasting of federal government proceedings. For cable interests, the initiative presented a low-cost way for cable to score points with federal regulators by performing public interest functions.[3]

For broadcasters, the trend toward cable program origination was troublesome. In written comments on the 1968–1969 proposed rules, the NAB warned of the "dangerously destructive potential" for cable to develop into a form of pay television that would siphon off viewers, advertising, and the best programming from free broadcasting.[4] If this occurred, the NAB argued, lower-income Americans would be deprived of high-quality entertainment and public service programming, a result which would undermine the FCC's longstanding vision for broadcasting. To avoid this, broadcasters wanted cable program origination to be voluntary, commercial free, and limited to public service programming on a single channel.[5] However, given the FCC's new vision for the future of cable, broadcasters had little hope of obtaining regulations that simultaneously froze distant-signal importation and severely limited cable program origination.

Cable interests, though anxious to enter program origination, nevertheless opposed regulations making it a formal requirement. The NCTA went so far as to openly question whether the FCC had the authority to impose origination requirements, arguing that such regulations were "hardly ancillary to the Commission's responsibilities for regulation of television broadcasting."[6] Furthermore, the NCTA objected to the FCC's efforts to impose public interest requirements on cable; the cable group accused the agency of making "the medium fit the regulation instead of making the regulation fit the medium."[7] The NCTA called on the FCC, therefore, to recognize cable as a distinct new medium with the potential to serve the public interest by providing consumer choice. When viewed in this way, the proper role for government was to help cable reach its potential—not to force it into a narrow regulatory box.

The Burch Commission

Throughout 1969, the cable industry faced an uphill battle in its efforts to move FCC policy in a new direction. For years, commissioners whose first priority was protecting established interests from economic harm had dominated the commission. However, prospects for change improved in the fall of 1969 when Dean Burch was confirmed as the new FCC chairman. A lawyer and a former Republican National Committee chairman, Burch (much like Minow) came to the FCC with virtually no experience as a regulator.[8] Unlike Minow, however, Burch did not come to the FCC with definite ideas regarding the future of television. When asked about the cable issue in his nomination hearing, Burch admitted that he had "no great familiarity" with the issues; he noted only that regulatory solutions should follow "the charge of the Communications Act that the United States wants the best communications system available."[9] Burch's words suggested the possibility of a fresh pragmatism at the commission—comments he made later in the hearing process seem to confirm this:

> I do not believe the Commission should ever irrevocably commit itself to a certain course of action. Communications would appear to be one of the most dynamic and changing fields, and it follows that the Commission must be alert to revise its regulatory policies, as technology changes.[10]

This willingness to adjust and change course certainly became evident in Burch's approach to cable television policy.

Quite early on in his tenure, Burch signaled a desire to make cable a top priority. In January of 1970, the CATV Task Force was transformed into the Cable Television Bureau and thus became a permanent agency office on an equal footing with the Broadcast and Common Carrier Bureaus. By May, Burch had obtained a tentative agreement among a majority of the commissioners regarding a new set of regulations designed to promote cable growth. Burch presided over an eclectic commission staffed by a mix of liberal Democrats and conservative Republicans, and members of both parties were sympathetic to broadcasting interests. Nevertheless, Burch was more than willing to compromise and adjust policy proposals to win over even the most vocal critics.[11]

Early in the cable proceedings, Burch managed to win the vote of Nicholas Johnson, an outspoken Democrat on the commission. Although Burch and Johnson clashed on many issues, they both shared the general view that government needed to promote, not limit, cable's potential. Of the two men, Johnson perhaps had the more developed vision for cable. He advocated mandatory forty-channel systems in the largest cities, with half the channels set aside for broadcast station carriage and the other half set aside for program origination. Johnson believed that space should be set aside among the origination channels for educational, governmental, and general public access use. Even though Burch's initial cable proposal in 1970 did not fulfill all of Johnson's goals, Johnson tentatively supported it. He later explained his vote, noting that the proposal "was a compromise we agreed was feasible, and one under which cable could at least get started."[12]

Burch's decision to compromise with interest groups, however, eventually cost him the support of some commissioners. When his initial proposal failed amid interest group objections, Burch concluded that resolution of the cable controversy could occur only through a political compromise. Unable to engineer such a compromise within the legal and practical constraints imposed upon an independent regulatory commission, Burch turned negotiations over to the White House Office of Telecommunications Policy (OTP). The resulting "consensus agreement" was then returned to the FCC as a fait accompli, and no additional input was taken from either the commissioners or the public. The OTP negotiation process prompted a strong written dissent from Commissioner Johnson, who accused the chairman of engaging in "secret bargaining sessions" that made "a mockery of the FCC's independence and role as an arm of Congress."[13] Commissioner Robert E. Lee, a longtime supporter of UHF television, also dissented, contending that the compromise excessively favored cable at the expense of broadcasters. Nevertheless, by 1972, Burch had achieved his goal of developing a broad new set of cable regulations.

The First Steps, 1970–1971

The first steps toward comprehensive cable regulation were taken in early 1970 when Burch instructed general counsel Henry Geller to develop a working policy proposal. Although Geller had been the author of more restrictive regulations under previous chairmen, the "Public Dividend Plan" (as the proposal later came to be known) reflected Burch's desire to more evenly balance cable growth with the economic protection of other interests. By May 1970, the proposal was circulated to the commissioners, who quickly voted to undertake a rulemaking proceeding.[14]

The Public Dividend Plan proposed to allow cable systems in the largest one hundred markets to import four distant signals into their service areas. In exchange, cable systems would be required to delete all of the imported commercials and replace them with advertising sold by local broadcast stations. Cable systems could also import an unlimited number of non-commercial, educational distant signals, as long as those stations consented to carriage. In exchange for the signal carriage privileges, cable systems would agree to make copyright payments according to a schedule set by Congress; in addition, cable systems would contribute 5 percent of their revenue to support the Corporation for Public Broadcasting. The commission also sought public comment on whether cable systems should be required to provide channels for local government, educational, and public access use, as well as leased channel space for third party commercial programming.[15]

In theory, the Public Dividend Plan contained something for everyone. Cable systems obtained liberal distant-signal importation provisions, while broadcasters gained advertising revenue. Copyright owners gained some compensation, and Commissioner Johnson and other advocates for the public interest gained support for educational television and new outlets for local expression. Nevertheless, the proposal garnered very little support. Broadcasters opposed virtually all parts of the plan. The NAB denounced the distant-signal provisions as a "drastic departure" from the commission's established goals of protecting and promoting broadcasting,[16] and echoing this sentiment, the AMST highlighted deficiencies in the commercial substitution provisions, which it viewed as technically and economically unworkable. Furthermore, AMST argued, the proposal was "philosophically wrong" because it allowed broadcasters to make money by selling advertising time on imported signals, fragmenting their own audiences in the process.[17]

Cable operators were only moderately more supportive of the proposal. The NCTA supported the commercial substitution portion of the plan, but some mid-sized cable providers, such as Midwest Video, viewed the scheme as costly and challenging to implement, both technically and administratively. Virtually all cable operators were unenthusiastic about the educational television subsidies and generally questioned whether the commission had the authority to impose these requirements. Cable operators generally supported the notion of copyright payments, although they cautioned that any payment structure must not be so rigid as to financially undermine smaller providers.[18]

The Public Dividend Plan was an attempt at compromise rather than a set of measures rooted in any extensive research or practical knowledge possessed by the FCC. As a result, the commission was in no position to challenge the industry's contentions regarding the plan's feasibility. Moreover, the commission was unwilling to implement such a complex plan with virtually no industry cooperation. The Public Dividend plan was essentially dead by early 1971, and the FCC was left casting about for new policy alternatives.

Seeking Clarity, 1971

The FCC's search for policy alternatives continued to be complicated by a lack of reliable information concerning the economic impact of cable on broadcasting. Although new data gradually became available, studies were often contradictory—a fact which only exacerbated the prevailing sense of ambiguity and did nothing to point the way to clear policy solutions. In 1970 RAND produced what was perhaps the most robust economic study up to that point. Using data from the 1960s as a baseline, this study employed a computer model designed to predict the economic impact of cable through the 1980s. Contradicting the longstanding claims of broadcasters, the RAND study predicted that cable had a positive rather than a negative economic effect on UHF broadcasting: by improving UHF audience reach and picture quality, cable eliminated the competitive advantage long held by

VHF stations. If anything, VHF suffered greater revenue losses from cable carriage of UHF competitors than the UHF broadcasters themselves did. VHF stations, however, tended to be network affiliates with larger profits and could suffer moderate revenue losses without experiencing a severe impact on programming or service. Ultimately, the study concluded that the largest revenue losses occurred in smaller television markets, suggesting that the FCC's focus on protecting urban broadcasters was misplaced.[19]

Although the 1970 RAND study was more sophisticated than previous analyses, it was an effort to predict *future* trends through a computer simulation. Broadcasters, by contrast, produced data suggesting a substantial *present* impact from cable television. The NAB commissioned a study that estimated an incremental decrease in broadcasters' prime time audiences with each distant signal imported into large markets by cable operators.[20] These conclusions were supported by a second study—commissioned by a group of twenty broadcasters—which suggested that many UHF stations were still of marginal profitability and remained vulnerable to the effects of cable audience fragmentation. In essence, broadcasters contended that little had changed since the early 1960s, when the entire focus of cable regulation had been the protection of UHF broadcasting.[21]

The FCC's own capacity to analyze trends in the cable industry was still weak. The Cable Television Bureau remained a small office, the staff of which was mostly dedicated to routine implementation responsibilities rather than policy analysis. Although the Broadcast Bureau produced an analysis of cable's economic impacts in 1970, the conclusions were tenuous and did little to resolve the analytical gap between the RAND report and the broadcaster-backed studies.[22] As had been the case in the 1960s, however, a lack of analytical clarity did not lead to policy paralysis at the FCC. Burch and his fellow commissioners once again sought out a compromise that would both promote cable development and provide some protection for broadcasters and copyright owners.

In the spring of 1971, the FCC began its search for a compromise solution by organizing a series of panel discussions involving a wide range of stakeholders in the cable television debate. Somewhat informal in their organization, the sessions differed in character from the oral arguments typically used in FCC proceedings. In his opening remarks, Chairman Burch described the sessions as "an experiment designed to improve the process by which we acquire information."[23] More importantly, the sessions were an effort to foster dialog between contending interests in an attempt to reach some common ground. Burch made clear, however, that progress toward resolution was only possible if the participants moved beyond entrenched positions.[24]

The discussion sessions produced a particularly constructive dialog regarding the distant-signal importation issue. Cable interests argued that distant-signal importation provided the "economic base" that was needed to produce growth in program origination. For perhaps the first time in any FCC cable proceeding, major broadcasting interest groups seemed moderately receptive to this idea. The NAB and the AMST both endorsed a "3–1–1" plan. In essence, they argued that "full" television service constituted three network signals, one independent signal, and one non-commercial signal. When a service area lacked any or all of these components, cable would be permitted to import enough distant signals to fill in the gaps. In some instances, broadcasters even seemed open to distant-signal importation in urban areas, where growth in cable program origination was most viable.[25]

This shift in broadcasters' views concerning distant-signal limitations reflected a belief that these regulations would provide them with modest (at best) economic protection in an environment characterized by cable program origination. Broadcasters therefore favored broadened exclusivity rules designed to limit cable's ability to compete with broadcasters for commercially produced program content. Several years earlier in the *Subscription TV* proceeding, the FCC had designed a series of "anti-siphoning" rules that limited the ability of over-the-air pay television channels to gain access to feature films, sporting events, and series programming.[26] In a 1970 order, the FCC extended these rules almost verbatim to apply to pay-per-view cable programming as well.[27]

As a result of the panel sessions, the major points of contention became clearer in the minds of the FCC commissioners. The task now was to translate this new information into workable policy. As one high-ranking FCC staffer remarked to a prominent trade publication, the process of arriving at this balance would be "political, not logical;" it would be "political in the highest sense of the word" and characterized by "compromises to accommodate the contending factions."[28] Indeed, the process that unfolded over the next several months was nothing short of an exercise in the art of compromise.

FORGING POLICY, 1971–1972

With the panel sessions now complete, the commission once again asked staff to formulate working policy proposals. Two options papers, prepared by the FCC's Planning Office and the Cable Television Bureau respectively, reflect the growing consensus concerning the distant-signal importation issue. Although they differ on the exact details, both papers support liberalization of the large-market distant-signal rules, and the Planning Office paper favors a 3–1–1 approach similar to that suggested by broadcasters in the panel sessions. Both papers acknowledge the need for cable to serve as an outlet for local expression, with the Planning Office's proposal emphasizing the need for educational, governmental, and public access channels. Whereas both proposals recognize the need for resolution of the copyright issue, the documents indicate a growing consensus that this was a legislative matter that ought to be separated from the FCC's rulemaking endeavors.[29]

With the staff recommendations in hand, Burch asked Cable Television Bureau chief Sol Schildhause to work on a formal proposal for a comprehensive set of cable television regulations.[30] Making good on a promise to keep Senate Commerce Committee chairman John Pastore informed of the FCC's activities, Burch first publicly released the finalized proposal as a "letter of intent" to Congress in August of 1971. The proposal covers a broad range of regulatory matters, including public access issues, technical standards, and the federal-state division of authority over cable. Perhaps the letter's most important contribution, however, is the outline of a program that Burch hoped would finally resolve the contentious signal carriage issue.[31]

As a part of the proposal, the basic 1965–1966 must-carry and network non-duplication requirements were to be maintained. In addition, the FCC defined the "minimum" level of television service an area should be able to support based on its market size. For the largest fifty markets, minimum service was defined as three network and three independent signals. Markets fifty-one to one hundred were expected to support three networks and two independent signals; for smaller markets, the requirement was three networks and one independent signal. Where communities were not receiving minimum television service through local broadcast outlets, cable operators would be allowed to import signals to fill in the gaps. In addition, cable operators in the largest one hundred markets would be permitted to supplement minimum service with two additional distant signals, subject to certain "leapfrogging" restrictions that required them to favor nearby UHF independents and signals within their states.[32]

These complex signal importation limitations served the dual goals of providing cable systems with increased freedom to expand into urban areas while also providing some protection for UHF broadcasters. Broadcasters were further protected by an extension of the exclusivity restrictions that the FCC proposed to place on sporting events. When importing sporting events on distant signals, cable systems were required to observe the "blackouts" that were mandated in league-television agreements.[33]

The letter of intent was initially approved by the FCC with a vote of 6–1. Broadcasters, however, quickly indicated that they were dissatisfied with the signal importation and exclusivity provisions; this expression of discontent led to doubts among some commissioners and created a need for further compromise. To help engineer consent, Burch turned to Clay Whitehead, director of the White House OTP. At the time, Whitehead was in the process of negotiating a copyright settlement among broadcasters, cable interests, and copyright owners; the OTP provided an environment in which industry groups were already engaged in ongoing talks. Furthermore, because it was not a regulatory body, the OTP could negotiate in a more closed-door atmosphere without procedural or legal restraints. In November of 1971 and with Burch's approval, Whitehead negotiated a consensus agreement to resolve the signal importation and exclusivity disputes.[34]

The Consensus Agreement and its Aftermath

The consensus agreement negotiated by the OTP made several key changes to the letter of intent. Exclusivity protections were extended to cover syndicated series programming imported through distant signals into major markets. In addition, the leapfrogging provisions were altered slightly to favor importation of the closest distant signals from the largest twenty-five markets, and all parties agreed to support copyright legislation.[35] The changes were formally ratified by the FCC in 1972 as part of a comprehensive package of cable regulations, the major points of which can be summarized as follows:

- Cable systems were required to carry all local broadcast signals.
- Cable systems were required to provide minimum television service to all television markets. If necessary, the cable system could import distant signals to bring the market up to minimum service.
- Cable systems in the largest one hundred markets could import up to two additional distant signals.
- Leapfrogging rules required that signals imported from the largest twenty-five markets be from the closest such markets.
- Non-duplication of network programming was carried over from the 1966 rules. Cable systems could not duplicate network programming on the same day it was aired by a broadcaster in the service area.
- Syndicated exclusivity rules guaranteed non-duplication of non-network programming, particularly sporting events and series programming. Anti-siphoning rules also limited cable's ability to use feature films and sporting events as pay-per-view programming.
- Program origination requirements, first proposed in 1968, were formally applied to cable systems with over 3,500 subscribers as of 1971. Cable program origination was to be commercial free and subject to public interest regulations.

- Access and local expression demands were fulfilled through rules requiring cable systems to maintain free public, educational, and governmental channels, and lease access to unused channels on a common carrier basis.
- Channel capacity for all new cable systems was set at a minimum of twenty channels.

CONCLUSION

In 1972—after a decade of experimentation, negotiation, and court battles—the FCC finally put in place a comprehensive scheme of cable television regulations. The ultimate road to the scheme's adoption, however, was far from orthodox. Rather than functioning as the kind of expert commission envisioned by the Progressives, the Burch FCC had employed bargaining and compromise as their methods of policy formulation. Furthermore, considered in their entirety, the 1972 rules resemble a comprehensive piece of legislation rather than a disparate collection of narrow administrative regulations. This is an impressive feat given that the FCC drew its authority from a statue that did not once include the phrase "cable television." The road to the 1972 cable rules was unique in FCC history and rare in the history of US administrative law.

The exceptional nature of the 1970–1972 rulemaking process was perhaps inevitable considering that resolution of the cable television problem rested squarely on the FCC's shoulders. With Congress unable to act, interest groups at loggerheads, and the courts providing little recourse for broadcasters and copyright holders, action on cable issues could only have come through administrative fiat. In taking action, however, the FCC was working with a limited knowledge base derived mainly from its experiences with the 1965–1966 rules and a handful of contradictory economic studies. For this reason, the FCC once again committed to learning by doing; it allowed a range of interests and ideas to be stated, worked out a compromise, and generated knowledge through practical experience over time. Indeed, upon issuing the 1972 rules, the agency expressed the belief that "experience and insight" would allow it to "act accordingly—to make revisions, major or minor—and to keep pace with the future of this dynamic area of communications technology."[36] Thus, the rules were a meant to be a major first step, but by no means the last step, toward finding a workable approach to cable television regulation.

ENDNOTES

1. *Broadcasting*, "Chance for Accord on Cable's Future?," 76, no. 22 (1969): 23.
2. Ibid., 24; *Broadcasting*, "Thumbs Down on the Cable Agreement," 76, no. 23 (1969): 25; *Broadcasting*, "Wrecking Crews Hit Cable Compromise," 76, no. 24 (1969): 45; *Broadcasting*, "Heat in the Ashes of a Doused Deal," 76, no. 26 (1969): 68.
3. *Broadcasting*, "Scratching the Itch to Originate," 76, no. 26 (1969): 69–70; *Broadcasting*, "Cable Opts for Networking Role," 76, no. 26 (1969): 71.
4. National Association of Broadcasters, Comments Relative to Paragraphs 11–20 and 23–25 of Part III of the Notice of Proposed Rulemaking and Notice of Inquiry, 3 June 1969, Docket 18397 (FCC Docketed Case Files, RG 173; Washington, DC: National Archives), 3.
5. Ibid., 4; See also Association of Maximum Service Telecasters, Comments, 3 April 1969, Docket 18397 (FCC Docketed Case Files, RG 173; Washington, DC: National Archives).
6. Here, the NCTA was referencing the US Supreme Court's tentative affirmation of FCC jurisdiction over cable in the *Southwestern* case. See National Cable Television Association, Comments on Paragraphs 11–20 and 23–25 of Part III of the Notice of Proposed Rulemaking and Notice of Inquiry, 3 April 1969, Docket 18397 (FCC Docketed Case Files, RG 173; Washington, DC: National Archives), 10–11.
7. Ibid., 13.
8. Kimberly A. Zarkin and Michael J. Zarkin, *The Federal Communications Commission: Front Line in the Culture and Regulation Wars* (Westport, CT: Greenwood Press, 2006), 162–163.
9. Senate Committee on Commerce, *Nominations—1969*, 91st Cong., 1st Sess., 1969, 5.
10. Ibid., 27.
11. *Broadcasting*, "In Sight: Stiffest Fight Yet Over Cable," 78, no. 21 (1970): 21.

12. *Cable Television Report and Order*, 36 FCC 2d 141, 310 (1972), Nicholas Johnson's Dissent.
13. Ibid., 306.
14. "In Sight," 21.
15. *Second Further Notice of Proposed Rulemaking in Docket 18397-A*, 24 FCC 2d 580, 582–587 (1970).
16. National Association of Broadcasters, Comments, 7 December 1970, Docket 18397-A (FCC Docketed Case Files, RG 173; Washington, DC: National Archives), 2.
17. Association of Maximum Service Telecasters, Comments, 7 December 1970, Docket 18397-A (FCC Docketed Case Files, RG 173; Washington, DC: National Archives), 25.
18. National Cable Television Association, Comments, 7 December 1970, Docket 18397-A (FCC Docketed Case Files, RG 173; Washington, DC: National Archives); Midwest Video, Comments, 22 October 1970, Docket 18397-A (FCC Docketed Case Files, RG 173; Washington, DC: National Archives).
19. Rolla Edward Park, *Potential Impact of Cable Growth on Television Broadcasting* (Santa Monica, CA: RAND, 1970).
20. National Association of Broadcasters, Comments in Docket 18397-A, 12.
21. Leonard L. Fischman and Associates, *Evaluation of FCC August 5, 1971 Distant-Signal Proposals for Cable Television in Terms of Their Impact on Over-the-Air Broadcasting*, 14 October 1971, Docket 18397-A (FCC Docketed Case Files, RG 173; Washington, DC: National Archives).
22. *Cable Television Report*, 168.
23. US Federal Communications Commission, Panel Discussion, 11 March 1971, Docket 18397-A (FCC Docketed Case Files, RG 173; Washington, DC: National Archives), 6.
24. Ibid., 5.
25. *Broadcasting*, "Where All the Talk about Cable May Lead," 80, no. 13 (1971): 82.
26. *Subscription TV Report and Order*, 15 FCC 2d 466, 556–565 (1968).
27. *Memorandum Opinion and Order in Docket 18397*, 22 FCC 2d 825 (1970).

28. "Where all the Talk," 82.
29. *Broadcasting*, "Two Blueprints for Unshackling CATV," 80, no. 17 (1971): 21–22.
30. Sol Schildhause, Oral History (Denver, CO: The Cable Center).
31. The letter of intent is reprinted as an appendix to the final cable television rules. See *Cable Television Report* 36 FCC 2d 141, 260 (1972).
32. Ibid., 263–266.
33. Ibid., 267.
34. Richard Olin Berner, *Constraints on the Regulatory Process: A Case Study of Regulation of Cable Television* (Cambridge, MA: Harvard University, 1975), 57–58; Don R. LeDuc, *Cable Television and the FCC: A Crisis in Media Control* (Philadelphia: Temple University Press, 1973), 198–199.
35. The consensus agreement is reprinted at the end of the 1972 Cable Television Report at 36 FCC 2d 141, 284 (1972).
36. *Cable Television Report*, 167.

Chapter 7

Re-Regulation to Deregulation, 1973–1980

The years from 1973 to 1980 constituted a period of marked change in the FCC's cable television policies. During these years, virtually all of the regulations that had been put in place in 1972 to protect broadcasters were abandoned. Though change initially proceeded incrementally, by 1979 the FCC was poised to abandon two of its most significant regulatory protections: the syndicated exclusivity and distant-signal carriage rules. As the FCC went through the process of reconsidering regulations in the mid to late 1970s, policy outcomes were shaped by knowledge rather than raw political compromise. The FCC at this time demonstrated a willingness to reflect on experience and to base policy judgments on pragmatic assessments of new information. These changes suggest that the commission had improved its ability to learn about cable television policy problems.

What factors explain the FCC's improved learning capacity? The political salience of cable television issues was certainly one consideration; in the mid-1970s, Congress began pressuring the FCC to reexamine some of the more restrictive cable regulations. Although political salience may have been an impetus for action, it nevertheless did not dictate regulatory outcomes. Rather, the activities of the FCC leaders fostered both organizational change and policy change. Chairmen Richard Wiley and Charles Ferris both believed that regulation should only be imposed if it was demonstrably in the public interest. Assessing the need for regulation, however, required the organizational capacity to analyze problems, sift through ambiguous data, and reach informed decisions. Fortunately for Wiley and Ferris, they were surrounded by a group of commissioners who shared their views and by a growing professional staff capable of providing expert analysis. Informed by an accumulated wealth of experience and empirical data, these actors gradually gained confidence that deregulation was the appropriate policy strategy.

INCREASING ORGANIZATIONAL CAPACITY

Among Dean Burch's most important legacies as FCC chairman was an increase in the agency's policy analysis and planning capabilities. For Burch, this meant not only increasing personnel but also adjusting the structure of the agency to improve the quality and flow of information. As Burch noted in a 1970 congressional hearing, the FCC's existing bureaus were busy with implementation tasks and rarely had the time or resources to focus on analysis.[1] To remedy this, he created a special Policy Planning Office designed to give the commissioners independent advice on a broad range of communications issues. In 1973 the Planning Office was renamed the Office of Plans and Policy, and by the end of the decade it had a staff of forty full-time employees.[2]

Burch also presided over an effort to increase the analytical capacities within the Cable Television Bureau. In 1973 the bureau was reorganized into four staff offices: two devoted to implementation matters and two devoted to analysis and policy development. Between 1972 and 1974, the bureau's staff was increased from thirty-two to ninety full-time employees.[3] As Cable Television Bureau chief Sol Schildhause later noted, by 1973 the bureau was becoming "a factor in the Commission."[4] When Schildhause left the FCC that year, he was replaced by David Kinley—an organizational outsider brought in by Burch to provide the bureau with a fresh perspective. Like Burch, Kinley took a pragmatic approach to regulation, preferring cautious, reasoned action to sudden changes in policy direction.[5] By the time Kinley left the FCC in 1976, he had helped transform what had once been a small, temporary staff office into a sizeable bureau with a reputation for effectiveness.

Furthermore, as the size of the commission's staff grew, new ideas and perspectives entered the commission. A growing number of staff believed overly restrictive regulations had slowed the development of cable television. Many staff also sought to place a higher value on analysis and empirical evidence as the bases for policy judgments; although compromise and negotiation might have made sense at one time, years of practical experience with cable television had provided little evidence of a negative economic impact on broadcasters. Unless broadcasters could make an affirmative showing of harm, there was little rationale for protecting them through regulation.[6] These attitudes marked a significant departure from the approach taken in the 1960s, when staff like Kenneth Cox and Henry Geller had taken a more protective tack by choosing to regulate even in the absence of empirical evidence.

FREE-MARKET PRAGMATISM: THE WILEY AND FERRIS COMMISSIONS

Following Dean Burch's departure from the FCC in 1974, his pragmatic approach to regulation was carried forward by his two immediate successors, Richard Wiley (1974–1977) and Charles Ferris (1977–1981). Although Wiley was a Republican and Ferris a Democrat, both FCC chairmen shared a core belief that marketplace competition was preferable to regulation unless experience and evidence demonstrated otherwise. Neither Wiley nor Ferris was interested in competition for competition's sake, and neither believed that deregulation was the best course of action in all situations. Rather, under both chairmen, proposed and existing regulations were carefully considered to determine whether they constituted the best policy choices. For this reason, Wiley and Ferris' style of regulation is here referred to as "free-market pragmatism."[7]

Like Burch, Wiley was a loyal Republican who came to the FCC with little regulatory experience; as Burch had done, Wiley developed a reputation for consensus building, frequently engineering unanimous commission votes.[8] Wiley, however, developed his own distinct regulatory philosophy, which he referred to as "re-regulation" and defined simply as "better, more intelligent regulation." The idea was to periodically reexamine existing regulations to determine whether they remained valid in light of changing conditions. Upon joining the commission in 1972, Wiley spearheaded the creation of a staff working group called the "Re-Regulation Task Force," which was to examine the continuing relevance of radio regulations.[9] Over the next seven years, the activities of this task force led to numerous alterations in established FCC rules.

When he became FCC chairman in 1974, Wiley created another re-regulation task force to do the same for cable television.

Charles Ferris pursued an even more deregulatory agenda than his predecessor had, placing great faith in microeconomic theory as a tool for evaluating the public interest—specifically, in the idea that the public interest is served when goods and services reflect public wants and are provided efficiently and at the lowest cost. When regulation helps achieve these goals, it is valid; when it does not, free markets should be favored. In rulemaking documents produced by the Ferris commission, extensive applications of microeconomic theory are sometimes used to justify preferred courses of action.[10]

Furthermore, whereas Wiley's re-regulation had often involved piecemeal reexamination of individual regulations, Ferris was not afraid to revisit the fundamental policy questions behind the regulations. For instance, in the 1979 deregulation of radio inquiry, the Ferris commission sought to examine whether regulation of broadcast programming content—a fundamental consideration in FCC policy since the 1940s—was still in the public interest. The commission did not reject localism or content diversity as important ends, but it concluded that at a time when much of the public had access to multiple local radio and television outlets, these goals could best be achieved through market forces. In order to reach this conclusion, the FCC conducted a thorough analysis in which it reflected on the validity of past policy assumptions in light of changing social and economic conditions. A similar analytical pattern was followed in 1977 when the commission opened a proceeding to reexamine the assumptions underlying its cable television regulations.[11]

Wiley and Ferris both received support from several commissioners who shared their desire to periodically reexamine regulations. Glen Robinson, a law professor and advocate of free-market principles, made clear from the beginning his belief that "the FCC must be willing on appropriate occasion to reexamine even its most well-established policies ... I do not mean necessarily changing them. I mean reassessing them."[12] Upon his nomination to a seat on the commission in 1976, Joseph Fogarty expressed the belief that the FCC could "encourage the growth of cable through enlightened regulation;" this basic sentiment was shared by Margarita White, who also joined the agency at that time.[13]

Free-market pragmatism created an intellectual climate in which experience and new information were given thoughtful consideration in FCC policy deliberations. The new climate impacted the thinking even of some commissioners who had in the past been staunch advocates of cable regulation. Robert E. Lee, for example, who had served on the commission since the 1950s, gradually and reluctantly changed his views. Although he was an advocate of free-market principles in general, Lee had earned the nickname "Mr. UHF" in the 1950s because of his strong interest in promoting the development of television broadcasting. Lee had supported cable regulation in the 1960s, but by the 1970s he was willing to reconsider some rules after experience had demonstrated that they were no longer necessary. For instance, in 1976 Lee supported a repeal of the leapfrogging rules once he became convinced that they provided no meaningful protection for broadcasters.[14] Likewise, Commissioner James Quello, a strong supporter of broadcast interests in most contexts, concluded "warily—and with some trepidation" that cable operators should be freed from leapfrogging requirements.[15]

Program Origination Reconsidered

The first major action taken in conjunction with Wiley's re-regulation effort involved a reconsideration of the FCC's cable program origination requirements. Although the rules were originally scheduled to take effect in early 1971, they were quickly stayed pending the outcome of a court challenge.[16] The US Supreme Court ultimately upheld the program origination rules,[17] but the FCC did not immediately move forward with their implementation because few interested parties were enthusiastic about the rules. Cable providers found them too restrictive, while broadcasters believed they were not restrictive enough. Furthermore, the 1970 RAND study had painted a complex picture of cable program origination, suggesting that it was generally not profitable and could only work in large urban areas where systems could offset their losses through other revenue sources.[18] Having little practical experience with cable program origination, the FCC decided to monitor its development for a few years before moving ahead with regulation.

By 1974, the FCC was prepared to revisit the program origination rules "in light of experience and events" that were casting doubt on the regulations' validity.[19] Responses to a request for public comments revealed little evidence that mandatory origination was in the public interest. The NCTA did a survey of 658 cable systems and found that roughly 70 percent of systems with over 3,500 subscribers were engaged in voluntary program origination.[20] Some individual providers, however, found program origination unworkable even in the largest metropolitan areas. Viacom International, for example, claimed that its operating expenses for original local programming were three times the revenue it received. Likewise, Cox Cable Communications reported losses of roughly $200,000 over a three-year period for programming originated in the San Diego area. The general consensus among these groups was that local program origination was expensive and unlikely either to attract extensive advertising revenue or to receive subscriber support in the form of ratings.[21]

For their part, broadcasters provided only moderate opposition to the proposed rule change. Metromedia, Inc., opposed the rule change on principle, reminding the FCC that a "dedication to local programming [was] etched deeply in its history."[22] The group also questioned financial losses as a valid reason for terminating the rule, particularly given that UHF stations made little if any money on local live programming. Metromedia, however, was somewhat isolated in its opposition to the rule change; generally speaking, broadcasters made little effort to mount an opposition, and ABC even provided nominal support for a change in the rules.[23]

In the end, the FCC concluded that cable systems could adequately serve as an outlet for local expression through the provision of public access channels. Therefore, the rules were amended in 1976 to eliminate the origination requirements and replace them with a more basic requirement—namely, that systems of 3,500 subscribers or more maintain equipment for the production and presentation of local public access programming.[24]

Network Non-duplication Reconsidered

Around the same time that reconsideration of the program origination rules was initiated, the FCC also opened a proceeding to reexamine the network non-duplication requirements. One of the original components of the 1965–1966 rules package, network non-duplication, was carried forward in 1972 with minimal reconsideration. Within a few years, however, the rule had become the subject of complaints from broadcasters, cable operators, and cable customers alike, prompting the commission to ultimately label it a "constant source of aggravation and irritation."[25] The controversy stemmed in part from the difficulties associated with implementing non-duplication. The rule placed a burden on broadcasters to notify cable systems of the programs they wanted protected from duplication. Cable systems frequently complained that broadcasters did not notify them in a timely fashion, a circumstance which could (and often did) lead to complications in carrying out the requests. Cable subscribers, for their part, considered the programming blackouts they experienced during non-duplication periods a deprivation of the service for which they had paid.[26]

Furthermore, practical experience revealed that the rules had yielded some unintended consequences. As originally written, the rules required all cable systems operating within a broadcaster's reception area to guarantee non-duplication. As a result, cable operators were sometimes forced to black out overlapping signals from nearby markets that many residents could receive using a rooftop antenna. In these instances, non-duplication was illogical because it provided the local station with no meaningful protection from audience fragmentation.[27]

These considerations led the FCC to revisit the non-duplication requirements in 1974. Because the commission had little current information concerning the impact of the rules, it hoped that public comments would provide the needed evidence and insight. Comments from both cable and broadcasting interests, however, were filled with rhetorical appeals rather than hard data. The NCTA argued that the lessons of experience should be the FCC's guide, pointing out that during the previous decade, gradual relaxation of the non-duplication period (reduced from an original fifteen days to a prohibition only of simultaneous duplication) had yielded no adverse impacts for broadcasters. Furthermore, the NCTA called the entire rationale for non-duplication an "educated guess" since the FCC had in fact never studied whether the rules actually protected broadcasters.[28]

Broadcast interests, for their part, viewed any effort to repeal the rules as a drastic departure from the conventional wisdom in cable regulation. The AMST, for instance, argued that there was no public benefit in allowing cable operators to provide the same programming on multiple channels. Furthermore, given that the number of cable systems had doubled and the number of cable subscribers had increased by 540 percent since non-duplication was implemented, the promotion of cable growth could not be cited as justification for repealing the rule. Broadcasters, therefore, encouraged the commission to take a precautionary approach and keep the rules in place until there was clear evidence that they were no longer needed to protect local stations.[29]

Ultimately, the more extreme positions offered by both sides proved equally unconvincing. Instead of these suggestions, the FCC adopted a targeted set of revisions designed to remedy the practical problems with the rules. Non-duplication was limited to a narrow, fixed-mileage zone in order to remedy the overlapping-signal problem. In addition, cable systems were given the option of carrying the protected signal during the duplication-blackout period. Finally, cable systems with fewer than one thousand subscribers were exempted from the rules altogether because of their limited potential impact on broadcasters.

The Leapfrogging Rules Reconsidered

Although the idea of leapfrogging restrictions was first raised during FCC proceedings in the mid-1960s, no formal rules had been proposed prior to 1971. At that time, the FCC offered no systematic analysis to demonstrate that leapfrogging was a problem; it noted only that "these rules, while providing cable systems with some flexibility of choice, are also designed to give an expanded market to stations that might otherwise be passed over."[30] The FCC was concerned that cable systems in some areas would bypass struggling nearby UHF stations in favor of more distant signals.

The intent of the leapfrogging rules, then, was to guarantee that more independent television stations around the country would gain cable carriage, thus increasing their audience size and their profitability. By narrowing the geographical proximity of distant-signal carriage, the FCC also hoped to ensure that cable television service would remain relevant to local viewing interests.

Experience with implementation, however, revealed that the rules did not have the anticipated results. The rules required that only systems importing distant signals from the largest twenty-five television markets choose the closest such markets. In theory, this meant that cable systems in Miami could import signals from Washington, DC, but systems in parts of Pennsylvania could not because they were closer to Baltimore and New York City and therefore were required to prefer those markets.[31] Furthermore, proximity prompted many cable systems to choose out-of-state signals rather than signals from metropolitan areas within their own states. In such cases, the FCC recognized that in-state stations were more likely to be of local interest and granted numerous waivers to the rules.[32]

Taking a recommendation from its re-regulation task force, the FCC opened a proceeding in 1975 to reexamine the leapfrogging rules. Public comments once again revealed that broadcasters' and cable operators' viewpoints were on opposite ends of the spectrum. The NAB warned that by reconsidering the rules, the commission was "on the verge of undermining the local service foundation of television."[33] Repeal of the leapfrogging requirements, it was argued, would cause cable systems to import all of their independent signals from Chicago, New York, and Los Angeles, creating a narrow class of national "superstations" that would garner the vast majority of both viewership and revenue.[34] The solution, according to the NAB, was to strengthen the rules in a way that encouraged stations to import closer signals from smaller markets.

The NCTA, by contrast, argued that a requirement that cable operators carry closer signals because they were more likely to be of interest to local viewers was illogical. If those stations were in fact of interest to their subscribers, cable systems operating in a free-market setting would naturally be inclined to seek out the desired signals. Furthermore, the NCTA asserted, the creation of superstations was unlikely because the microwave transmission costs would be prohibitive for many cable systems. The NCTA did, however, present extensive data demonstrating that the leapfrogging rules caused cable to develop fastest in those areas where it was most cost effective to import signals from the largest twenty-five markets. The result was that metropolitan suburbs were receiving the benefits of cable while rural areas were being left behind. For the NCTA, the preferred course of action was to repeal the rules and allow cable to grow naturally in response to market forces.[35]

Ultimately, the deciding factor for the FCC was an internal study which demonstrated that cable signal carriage did not significantly increase a broadcaster's advertising revenue. Thus, although there might have been some occasional benefits to viewers from cable's carriage of closer stations, the identifiable benefits of the leapfrogging rules to broadcasters were minimal. In contrast, the potential costs were high, taking the form of an increased administrative burden, decreased choice for subscribers, and retarded cable growth in some parts of the country. With this cost-benefit calculation in hand, the FCC repealed the leapfrogging rules in 1976.[36]

Congress Weighs In

By 1976, Congress was scrutinizing the FCC's cable television regulatory activities more closely than it had in the past. A congressional staff study completed that year took a highly critical tone, accusing the commission of compromising with the industry rather than making impartial decisions based on a thorough review of the facts. The report called for an approach philosophically quite similar to that being taken by the Wiley commission: regulation only when it was demonstrably in the public interest. For the authors of the report, however, this meant that additional regulations needed to be reexamined, including the distant-signal carriage limitations and the syndicated exclusivity rules.[37]

Following the release of the report, the House Subcommittee on Communications held the most extensive set of congressional hearings ever undertaken with respect to cable television issues. Over the course of fifteen days, legislators heard testimony from an eclectic group of participants—including scholars, industry representatives, public interest groups, and present and past FCC commissioners. Armed with the staff report and conversant in recent economic studies of cable television, many members of Congress entered the hearings prepared to challenge the long-held assumption that regulation was necessary to protect broadcasters. Indeed, when the sitting FCC commissioners came to testify, subcommittee chairman Lionel Van Deerlin made clear that the session was an opportunity for Congress and FCC commissioners to "discuss … evidence which suggests that substantial revision of cable television regulation may be warranted."[38]

During the question and answer session, the tone was at times adversarial, as legislators pressed the commissioners to explain why they were acting so slowly to revise cable regulations. Chairman Wiley and the other commissioners resisted; they committed only to continual study of the matter and pledged to take action as needed.[39] The hearings did not ultimately lead to the development of any new cable legislation, but they highlighted the growing salience of cable television issues and suggested to the commissioners that there were minimal political obstacles to further deregulation.

Although Congress passed no broad-based regulatory legislation, it did succeed in resolving the longstanding cable television copyright dilemma. Under revisions to federal law passed in 1976, cable operators were granted compulsory licenses that, in exchange for set copyright fees, allowed them to carry all signals authorized by the FCC.[40] The copyright law revisions were significant because they rendered moot an important rationale for the FCC's exclusivity rules. Exclusivity rules could no longer be justified as a way to protect property interests in programming; their sole justification at this point rested on the contention that they protected the profitability of broadcasters by preventing audience fragmentation and revenue loss—a contention that had never been successfully demonstrated by broadcasters.

THE SECOND ECONOMIC INQUIRY

In 1977 the FCC began a second inquiry into the economic impact of cable television. When the agency began its first inquiry in 1958, there had been little available data concerning the economic effects of cable television on broadcasters. Nearly two decades later, the situation had changed: numerous studies had been conducted by think tanks and academic researchers over the previous seven years. Furthermore, the FCC at this point had the capacity not only to critically analyze existing studies but also to conduct high quality studies of its own. Perhaps most importantly, the FCC was able to draw on more than a decade of (largely unsuccessful) experience with cable regulation. Through a new economic inquiry, the commission hoped to determine once and for all whether extensive regulatory protection of broadcasters was warranted.

Through this inquiry, the FCC sought to uncover and critically assess its own assumptions. When it solicited public comments in

1977, the commission acknowledged that cable regulation rested on an "intuitive model," which had never been proven accurate. It was assumed that if cable caused any fractionalization of the broadcast audience, it would automatically lead to reduced revenue and lower-quality television service. Rather than continuing to accept this assumption, the FCC now proposed to treat it as a hypothesis that could be tested for accuracy.[41] In order to accomplish this, the FCC committed itself to carefully scrutinizing all available data. The research parameters were narrowly defined to include four specific considerations: (1) demand and cost factors in cable growth, (2) audience diversion, (3) the broadcast audience–revenue relationship, and (4) service to the public.[42] In setting these parameters, the FCC signaled its desire to infer causation rather than rely on general statistical trends or anecdotal evidence for its policy decisions. Even if cable had led to some loss of audience for broadcasters, regulation would only be justified if that loss harmed television service by creating a direct reduction in revenue.

For broadcast interests, who sought regulatory protection, this evidentiary burden was difficult to meet. The NAB commissioned two studies, one of which asserted that distant-signal importation caused substantial audience fractionalization that could lead to potentially devastating consequences for independent and smaller-market stations. The second NAB study provided evidence of a direct relationship between audience diversion and station revenue.[43] The NAB results, however, stood in isolation from the findings of numerous independent researchers who found only modest evidence of economic impact.[44] The FCC also commissioned a new study from RAND that seemed to rebut many of the NAB's contentions. The RAND report predicted a slight increase in audience reach for independent UHF stations, essentially reaffirming the findings of its 1970 study. Moderate audience diversion was predicted for VHF stations and network affiliates, with the largest losses appearing in smaller markets.[45]

Unwilling to rely solely on external studies, the commission requested that the Cable Television Bureau conduct analyses of its own. Five studies were done in conjunction with the inquiry, touching on such matters as audience diversion and the relationship between broadcast station revenue and viewership for both UHF and VHF.[46] One study was a statistical analysis of cable impact on broadcast audience shares in fifty-six television markets. The results paralleled the basic findings of the RAND study with respect to audience diversion: 83 percent of markets with UHF stations experienced either an increase or no decrease in audience share, while most VHF markets experienced a slight decrease.[47]

Virtually all of the studies examined in the inquiry suggested that some stations experienced audience and revenue losses from cable television. Ultimately, however, the FCC concluded that modest economic harm no longer justified extensive regulation. For one thing, the Ferris commission viewed these studies through the lens of microeconomic theory, according to which cable television served the public interest as measured by three factors. First, it maximized consumer welfare, offering people choices for which they were willing to pay. Second, it did not lead to distributional inequities that favored one segment of society over another. Third, through greater competition, it increased—rather than decreased—the quality of the television product.[48] Therefore, competition could not be restrained simply because one or a few broadcasters might experience reduced profits. Such was the normal course of marketplace competition.

These studies reinforced the FCC's findings from more than a decade of experience with regulation. In its report of the inquiry, the FCC reflected on its experience in a passage that warrants quoting at length:

> The questions which troubled us in the past … have narrowed considerably as we have gradually permitted and gained experience with, a competitive environment for video services. The worst of our fears … caused us to clamp down severely on the expansion of cable services in 1965, 1966, and 1968. Over the next few years we learned enough of the relationships between broadcasting and cable to dispel many of these fears, and in 1972 we relaxed some of the constraints upon the cable industry's ability to increase its service areas and its service offerings. Since then, experience has taught us a good deal more. We initiated this [inquiry] almost two years ago to assess whether careful analysis of existing and new economic evidence would corroborate our experiential appreciation of the cable-broadcasting relationship. As a result of our experience and of this inquiry, the uncertainty which has afflicted the decision-making process in the past has been satisfactorily eliminated … Our experience of twenty years and our analysis in this proceeding lead to the conclusion that continued regulatory intervention is not merely unnecessary, it is counterproductive.[49]

With experience and new information pointing the way, then, the FCC moved to eliminate two of the major remaining regulatory protections: the syndicated exclusivity rules and the distant-signal importation limits.

The Final Steps, 1979–1980

By 1979, repeal of the syndicated exclusivity rules seemed long overdue. The rules were complicated and restrictive, forcing cable systems in the largest one hundred markets to block out syndicated programming for one to two years after its initial release. In the largest fifty markets, the rules applied even when the programming was not being carried by local stations. The rules might have made sense at a time when broadcasters and program suppliers sought to protect their property rights, but with the 1976 copyright law revisions in place, this rationale was no longer persuasive.

Years of experience had also demonstrated that the syndicated exclusivity rules were rarely used in practice. The rules were only being implemented in thirteen of the largest fifty markets. Due to grandfathering provisions, roughly 90 percent of cable customers in the largest one hundred markets received their service from exempted systems. Few copyright holders actually requested coverage under the rules, and even the NAB noted that the rules were "largely meaningless" as a protection for broadcasters.[50] Furthermore, studies done in conjunction with the second economic inquiry predicted little long-term economic benefit to broadcasters from retaining the rules. These conclusions were definitive, and the syndicated exclusivity rules were easily repealed in 1980.[51]

On the same day, the distant-signal carriage limitations were also repealed, though with quite a bit more controversy. Broadcasters continued to protest the policy change, though to no avail. Armed with the evidence collected during its second economic inquiry, the FCC acted without hesitation. When the order to repeal the rules was issued, chairman Ferris proclaimed in a concurring statement that the FCC had finally "replaced regulatory intuition with regulatory analysis."[52]

Conclusion

By 1980, most of the regulations enacted in 1972 had been eliminated. Between 1974 and 1976, the FCC repealed the program origination requirements and the leapfrogging rules and narrowed the network non-duplication rules. Following the second economic inquiry, the syndicated exclusivity and distant-signal importation rules were eliminated. Additionally, the federal courts struck down the anti-siphoning rules and leased access channel requirements[53] as overextensions of the FCC's authority. The only major element of the 1972 rules still in place was the requirement that cable systems provide public, educational, and governmental access channels at the local level. In terms of economic protection for broadcasters, the only remaining rules were the must-carry and network non-duplication rules originally enacted in 1966.

Although the courts provided some jurisdictional boundaries and Congress applied some pressure, these actions at most helped construct the political and legal context within which decision-making took place. The most significant factor in the deregulation of cable television was the FCC's ability to reflect on its experience and incorporate new knowledge in the policymaking process. This newfound ability was a product of the efforts of pragmatic leaders who improved the FCC's ability to collect, analyze, and disseminate information throughout the organization. A more detailed assessment is provided in the concluding chapter, but it seems fair to state at this point that by 1980, the FCC had learned to regulate—or perhaps deregulate—cable television.

ENDNOTES

1. House Committee on Government Operations, *Reorganization Plan No. 1 of 1970*, 91st Cong., 1st Sess., 1970, 55.
2. US Federal Communications Commission, *Forty-Fifth Annual Report to Congress* (Washington, DC: GPO, 1979).
3. US Federal Communications Commission, *Thirty-Ninth Annual Report to Congress* (Washington, DC: GPO, 1973); US Federal Communications Commission, *Fortieth Annual Report to Congress* (Washington, DC: GPO, 1974).
4. Sol Schildhause, Oral History (Denver, CO: The Cable Center).
5. In reflecting on his approach to regulatory decision making, Kinley later noted that "if it's not clear what the right answer is … you are better off not implementing any answer at that point." See David Kinley, Oral History (Denver, CO: The Cable Center).
6. Dean M. Krugman, "FCC Commissioner, Legal Assistant, and Staff Perceptions of Cable TV," *Journalism Quarterly* 56 (1979): 3–8.
7. The label "free-market pragmatism" is also a useful way to distinguish Wiley and Ferris' style from the more ideological approach to marketplace competition taken in the 1980s by chairmen Mark Fowler and Dennis Patrick.
8. Kimberly A. Zarkin and Michael J. Zarkin, *The Federal Communications Commission: Front Line in the Culture and Regulation Wars* (Westport, CT: Greenwood Press, 2006), 203.
9. *Broadcasting*, "Making Life a Bit Easier: Re-Regulation Gets Underway," 83, no. 19 (1972): 19.
10. See *Deregulation of Radio*, 73 FCC 2d 457, 492–516 (1979).
11. *Report in Docket 21284*, 71 FCC 2d 632 (1979).
12. Senate Committee on Commerce, *Nominations June–October*, 93rd Cong., 2nd Sess., 1974, 31–32.
13. Senate Committee on Commerce, *Nominations July–September*, 94th Cong., 2nd Sess., 1976, 28, 33.
14. *Report and Order in Docket 20487*, 57 FCC 2d 625, 646 (1976).
15. Ibid., 647.

16. *Notice of Proposed Rulemaking and Notice of Inquiry in Docket 19988*, 46 FCC 2d 139, 141 (1974).
17. *US v. Midwest Video Corp.*, 406 US 649 (1972).
18. N. E. Feldman, *Cable Television: Opportunities and Problems in Local Program Origination*, 21 October 1970, Docket 18397-A (FCC Docketed Case Files, RG 173; Washington, DC: National Archives).
19. *Notice in Docket 19988*, 141.
20. National Cable Television Association, Comments, 24 May 1974, Docket 19988 (FCC Docketed Case Files, RG 173; Washington, DC: National Archives), 13.
21. Viacom International, Inc., Comments, 22 May 1974, Docket 19988 (FCC Docketed Case Files, RG 173; Washington, DC: National Archives), 2; Cox Cable Communications, Inc., Comments, 22 May 1974, Docket 19988 (FCC Docketed Case Files, RG 173; Washington, DC: National Archives), 4–6, 10.
22. Metromedia, Inc., Comments, 24 May 1974, Docket 19988 (FCC Docketed Case Files, RG 173; Washington, DC: National Archives), 4.
23. American Broadcasting Company, Comments. 24 May 1974, Docket 19988 (FCC Docketed Case Files, RG 173; Washington, DC: National Archives).
24. *Report and Order in Docket 19988*, 49 FCC 2d 1040, 1106 (1974).
25. *First Report and Order in Docket 19995*, 52 FCC 2d 519, 522 (1975).
26. Ibid., 522–523.
27. *Notice of Inquiry and Notice of Proposed Rulemaking in Docket 19995*, 46 FCC 2d 1164, 1168 (1974).
28. National Cable Television Association, Comments, Docket 19995 (FCC Docketed Case Files, RG 173; Washington, DC: National Archives), 2–3, 6.
29. Association of Maximum Service Telecasters, Comments, Docket 19995 (FCC Docketed Case Files, RG 173; Washington, DC: National Archives), 2–4.
30. The letter of intent is printed as an addendum to the 1972 Cable Television Report. See 36 FCC 2d 260, 265 (1972).

31. *Notice of Proposed Rulemaking in Docket 20487*, 53 FCC 2d 175, 178 (1975).
32. See *Commission on Cable Television of the State of New York*, 43 FCC 2d 826 (1973); *Madison County Cablevision*, 42 FCC 2d 969 (1973); *First Illinois Cable TV*, 45 FCC 2d 304 (1974).
33. National Association of Broadcasters, Comments, 29 July 1975, Docket 20487 (FCC Docketed Case Files, RG 173; Washington, DC: National Archives), 8.
34. Ibid., 15. Evidently, the NAB reasoned that this would occur because of the superior sports programming televised in these areas.
35. National Cable Television Association, Comments, 30 July 1975, Docket 20487 (FCC Docketed Case Files, RG 173; Washington, DC: National Archives).
36. *Report and Order in Docket 20487*, 57 FCC 2d 625 (1976).
37. House Committee on Interstate and Foreign Commerce, Subcommittee on Communications, *Cable Television: Promise versus Regulatory Performance*, 94th Cong., 2nd Sess., 1976, Committee Print, 2–6, 39, 50.
38. House Subcommittee on Communications, *Cable Television Regulation Oversight—Part II*, 94th Cong., 2nd Sess., 1976, 1229.
39. Ibid. For instance, see the exchanges between Chairman Wiley and Rep. Tim Wirth at 1261–1264.
40. Stanley M. Besen, Willard G. Manning, Jr., and Bridger M. Mitchell, "Copyright Liability for Cable Television: Compulsory Licensing and the Coase Theorem," *Journal of Law and Economics* 21 (1978): 68–70.
41. *Notice of Inquiry in Docket 21284*, 65 FCC 2d 9 (1977).
42. Ibid., 17.
43. Wharton EFA, Inc., *Impact of Cable TV on Local Station Audience*, March 1978, Docket 21284 (FCC Docketed Case Files, RG 173; Washington, DC: National Archives); Charles River Associates, Inc., *The Audience Revenue Relationship for Local Television Stations*, 4 December 1978, Docket 21284 (FCC Docketed Case Files, RG 173; Washington, DC: National Archives).
44. *Report in Docket 21284*, 668–674.
45. Rolla Edward Park, *Audience Diversion Due to Cable Television: A Statistical Analysis* (Santa Monica, CA: RAND, 1979), 1–2.

46. US Federal Communications Commission, Public Notice: Cable Television Bureau Releases Research Studies Relevant to Cable Inquiry, Docket 21284 (FCC Docketed Case Files, RG 173; Washington, DC: National Archives).
47. US Federal Communications Commission, Cable Television Bureau, *Impact of Cable Television on Audience Shares*, Docket 21284 (FCC Docketed Case Files, RG 173; Washington, DC: National Archives), 1–2.
48. *Report in Docket 21284*, 636–640.
49. Ibid., 658–659.
50. *Report in Docket 20988*, 71 FCC 2d 951, 953 (1979).
51. *Report and Order in Dockets 20988 and 21284*, 79 FCC 2d 663 (1980).
52. Ibid., 891.
53. *Home Box Office v. FCC*, 567 F.2d 9 (1977); *FCC v. Midwest Video Corp.*, 440 US 689 (1979).

Chapter 8

Conclusion

This study is structured around the general thesis that organizational learning impacted the direction of FCC cable television regulation between 1952 and 1980. In this chapter, the thesis is assessed in light of the empirical evidence presented in chapters 3–7. Overall, the evidence supports the thesis, though it appears that learning was substantial at some times and noticeably absent at others. Although the FCC struggled to learn throughout the 1950s and '60s, learning became a particularly important factor in the 1970s as the agency considered proposals to deregulate cable television. The pages that follow provide a detailed evaluation of the learning thesis, and the study concludes with a brief account of regulatory efforts to protect broadcasters from cable television after 1980 and some final thoughts concerning the value of characterizing regulatory policymaking as a learning process.

Learning and Cable TV Regulation

Recall that organizational learning is generally indicated when a policymaking organization employs experience and other sources of knowledge in an effort to fashion more effective policy (see chapter 2 for full discussion). Learning may be conceptualized in terms of three orders of complexity, all of which are most likely to take place when the proper learning conditions exist. Building on these theoretical precepts, four research questions serve as guideposts for this analysis:

1. Did the FCC engage in first-order learning?
2. Did the FCC engage in second-order learning?
3. Did the FCC engage in third-order learning?
4. Did learning conditions advance or inhibit organizational learning?

These research questions are considered here in light of the empirical evidence.

First-Order Learning

Was FCC cable television regulation influenced by first-order learning during the period under study? The answer appears to be a qualified yes. First-order learning was substantial, though by no means uniform, throughout the period under study. The first two decades were marked by a distinct absence of first-order learning owing in no small part to a lack of knowledge about cable television. Cable television presented the FCC with a highly novel policy problem that proved difficult to resolve within the context of existing ideas and regulatory models. Because the FCC was unable to draw upon an existing knowledge base for solutions, regulatory decision making became a contested dialog involving FCC commissioners and staff, interest groups, and external experts—all of whom offered competing issue frames, policy analyses, and solutions. The ideas put forward by these parties were frequently ambiguous and contradictory, leaving the FCC to function in a highly uncertain decision-making environment.

With no clear, unified voice pointing the way to solutions, commissioners used their regulatory philosophies and instincts as guides. In the 1950s, the decision not to exert regulatory authority over cable reflected the conservative instincts of the Doerfer commission. With little new knowledge or experience to guide them, liberal commissioners in the 1960s actively pursued regulation—a choice that reflected their strong belief in the need to preserve television as a public interest enterprise. The early 1970s brought a somewhat more pragmatic group to the commission, but the stock of cable television knowledge and experience continued to increase slowly. As a result, the FCC turned to bargaining and compromise as a way to formulate a complicated and extensive set of cable regulations. The only notable first-order learning during this time occurred shortly after completion of the 1966 regulations, when experience demonstrated that existing implementation processes were unworkable. In this instance, however, learning was confined primarily to questions of administrative practicality and ultimately had little direct impact on the content of regulatory policies.

Over time, however, the processes of reaching both the 1966 and the 1972 rules allowed the FCC to learn by doing. Practical experience with policy implementation led FCC officials to conclude that broadcasters needed only a modest amount of regulation as protection against any economic harm that might be caused by cable television. By 1975, the agency's experience counseled against the imposition of mandatory cable program origination requirements and demonstrated that the leapfrogging rules were unnecessary. Likewise, roughly a decade of experience helped the FCC devise a more targeted, practical version of the network non-duplication rules. After nearly eight years of experience with the syndicated exclusivity rules, the FCC was ready to conclude in 1980 that these regulations, like the others, were no longer needed.

The FCC was able to utilize scientific research when practical experience alone provided an insufficient knowledge base. By the late 1970s, a growing body of research overwhelmingly supported the relaxation or elimination of regulations that had been designed to protect broadcasters from cable television. No longer faced with ambiguous and contradictory findings, the FCC believed it could, based on empirical evidence, justify elimination of the distant-signal carriage rules—one of the most important and controversial components of its cable television regulatory structure. Powerful broadcast interests produced studies highlighting the need for continued distant-signal limitations, but their claims were drowned out by the many competing studies conducted by academics, think tanks, and FCC staff.

In short, because the FCC actively used practical experience and scientific research to refashion policy techniques, first-order learning was substantial between 1975 and 1980.

Second-Order Learning

Second-order learning occurs when an organization reorients its core policy beliefs. The evidence presented in this study suggests that the FCC experienced a gradual but ultimately substantial reorientation of its policy beliefs beginning in 1968. Prior to that time, a combination of law and organizational practice had embedded in the FCC a shared understanding that telephony and broadcasting were two species of public interest industries, each of which was subject to its own distinct model of regulation. The advent of cable television caused something of an intellectual crisis for the agency because the new medium could not be easily analogized to existing communications industries or technologies. This became particularly problematic in the mid-1960s when the commission wished to regulate cable television and needed to justify its actions in light of existing statutory authority. Initially, the FCC avoided any need to alter its policy beliefs; it did so by treating cable television as nothing more than an ancillary component of the broadcasting system. According to this categorization, cable television was only useful as a medium for carrying broadcast signals to areas where there was no local television station. The FCC had no desire to allow cable to impede the growth of broadcast television, undermine existing stations, or interfere with the property rights of station owners and content suppliers. To prevent these potential problems, regulators imposed must-carry requirements, distant-signal importation limits, and non-duplication rules.

Beginning in 1968, however, regulatory decision making reflected a gradual shift in policy beliefs—a change brought about by two trends. First, cable television was a distinct medium capable of originating content, providing innovative new services, and serving as an outlet for local expression in urban areas. These ideas were initially articulated in government reports and think tank studies, and they quickly found their way into FCC working papers and policy proposals. In many respects, the new ideas were not a dramatic departure from existing policy beliefs, nor did the new position mitigate regulators' concern that cable television might cause economic harm to broadcasters. The fresh vision for cable television did, however, lead the FCC to recognize the need for policy change. Government regulation needed to both promote the development of cable television and protect the profitability of broadcasting. To these ends, non-duplication and exclusivity protections were extended, but distant-signal limitations were relaxed in an effort to promote cable growth. At the same time, cable was expected to serve local audiences by engaging in program origination and making public access, educational, and local government channels available.

A second trend impacting FCC policy beliefs was the gradual infiltration of free-market thinking into organizational practice. This process began in the early 1970s when the FCC increased the number of professional economists and policy analysts it employed. These new staff members were willing to shift the burden of proof to broadcasters, who were thus forced to affirmatively demonstrate that regulatory protection was necessary. The new staff's perspective found a voice under chairmen Richard Wiley and Charles Ferris, the latter of whom was particularly interested in using microeconomic theory to determine whether regulation was in the public interest. Through the application of microeconomic theory, the public interest became synonymous with consumer welfare. This is not to say that regulators dismissed traditional goals like localism and content diversity; rather, they assumed these values would best be supported through free-market competition unless systematic analysis demonstrated otherwise. The trend toward free-market thinking observed in this study supports the more general conclusion that in the 1970s, an influential new epistemic community was changing the conversation concerning government regulation.[1]

These two intellectual trends together altered the way regulators understood cable television. By the end of the 1970s, cable was conceptualized not as a component of the broadcast system but as a unique video programming medium. To the extent that cable could be held to public service responsibilities, these requirements would not take the same form as those imposed on broadcasters. Whereas broadcasters were held to local programming requirements and other content limitations such as the fairness doctrine, cable could best serve as an outlet for local expression through the provision of public, educational, and governmental channels. No longer could broadcasters expect to be economically protected from cable, any more than they could expect to be protected from each other. Broadcasters and cable service providers were now two elements in a competitive video programming marketplace.

This intellectual shift was not a fundamental reconstruction of policy beliefs. It was, however, a substantive reorientation of existing beliefs about the public interest in television. For this reason, it is reasonable to conclude that at least a moderate degree of second-order learning had occurred within the FCC by 1980.

Third-Order Learning

Did the FCC engage in third-order learning (learning to learn) between 1952 and 1980? Recall that third-order learning is indicated when policymaking organizations improve their policy analysis and problem-solving capacities and demonstrate the ability to reflect on past actions and assumptions in an effort to improve future policy practice. Judging by this standard, the evidence demonstrates that the FCC did, in fact, learn to learn during this time.

By 1980, the FCC had improved its ability to analyze and solve cable policy problems. This assertion is perhaps best illustrated through a comparison of the FCC's two economic inquiries. When the first inquiry began in 1958, there was a paucity of readily available data concerning the economic impact of cable television on broadcasters. When the inquiry was completed a year later, the FCC had done little to remedy the situation: the data supplied in the inquiry came almost exclusively from interest group pleadings and consisted mostly of anecdotal evidence. The FCC made almost no independent effort to gather data and conduct systematic policy analysis. This outcome reflected the FCC's lack of internal expertise regarding cable television matters.

Two decades later, the situation looked quite different. When the FCC began the second inquiry in 1977, there was an expansive array of sophisticated data and studies concerning the economic impact of cable television on broadcasters. Had the agency possessed only the limited analytical capabilities it had in 1959, it might easily have been bogged down in all of the data. The FCC of the late 1970s, however, was prepared for the task at hand. From the outset, the FCC articulated a clear set of research parameters for the study, signaling to interested parties that it favored carefully constructed empirical analysis over rhetorical appeals and anecdotal evidence. Possessing the analytical tools supplied by economic theory, FCC staff not only critically examined a range of external studies but also conducted analyses of their own. Overall, the level of analytical sophistication in the second inquiry was substantial when compared to that of the first.

Furthermore, the final report in the second economic inquiry demonstrates a willingness on the part of the FCC to engage in organizational self-reflection. The FCC sought to bring to light the "intuitive model" that had long guided its cable television policies. Reflecting on nearly two decades of experience, the agency concluded that past decisions had been based on limited information and faulty assumptions. Perhaps this conclusion was unavoidable given the overwhelming amount of empirical evidence counseling against continued regulation. Nevertheless, the FCC's willingness to wade through extensive data, reflect on relevant experience, and gather new knowledge marked a significant change from the decision-making processes of earlier years. These considerations all suggest that the FCC engaged in third-order learning.

Learning Conditions

Chapter two presents the theory that organizational learning is likely to occur when favorable learning conditions exist—that is, when four factors are in place: (1) pragmatic leadership, (2) organizational capacity, (3) a favorable institutional context, and (4) a favorable political context. The evidence reviewed in this study suggests that learning became a more important factor in FCC regulatory decision making during the 1970s. Thus, if the learning thesis is correct, then the 1970s was a time when conditions were particularly favorable to learning at the FCC. An examination of changes in each of the four factors can help determine whether FCC learning conditions did in fact become more favorable during that time.

First, organizational learning is most likely when leaders are pragmatic and open to experimentation with a range of policy options. Prior to the 1970s, these characteristics were not evident among FCC leaders. During the 1950s, John Doerfer and his colleagues favored free markets and a very narrow reading of the Communications Act, and as a result they were reluctant to impose regulation on cable television unless specifically authorized to do so by Congress. Although the Doerfer commission did request legislation granting such authority, it was extremely narrow in scope and left virtually no room for regulatory experimentation. In the 1960s, on the other hand, the FCC was dominated by leaders who sought to regulate cable television and believed the Communications Act gave them adequate authority to do so. However, these commissioners were also deeply committed to the preservation of broadcasting as a public service enterprise. Ultimately, their intellectual commitments led them to impose restrictive regulations on cable television at a time when external studies were beginning to recognize the vast potential of the new medium.

By contrast, pragmatism was evident among FCC leaders throughout the 1970s. Pragmatic leadership first became apparent under the chairmanship of Dean Burch, who made the promotion of cable television a top priority early in 1970. In order to accomplish his goals, Burch was willing to consider a range of policy options and negotiate with industry leaders to reach a consensus. Burch's strategy resulted in the 1972 cable rules, which set the stage for future learning by the FCC. However, it was under Burch's successor, Richard Wiley, that learning first figured prominently into FCC decision making. The Wiley commission exhibited a regulatory philosophy referred to in this study as free-market pragmatism. Wiley and his colleagues favored competition unless regulatory intervention was demonstrably in the public interest. Thus, they were open to the possibility of continuing some regulation, but only when experience and empirical evidence counseled in favor of doing so. Employing this philosophy, the Wiley commission repealed the cable program origination requirements and leapfrogging rules and significantly narrowed the network non-duplication regulations. Chairman Charles Ferris, with his preference for microeconomic analysis, was perhaps less pragmatic than his predecessors, but he was still open to basing regulatory decisions on thoughtful consideration of experience and evidence. The Ferris commission spearheaded the second economic inquiry and ultimately chose to repeal the distant-signal carriage and syndicated exclusivity rules (while leaving the must-carry and network non-duplication rules in place).

As predicted, pragmatic leadership also coincided with an increase in organizational capacity. In 1970 chairman Burch created the Cable Television Bureau, signaling a desire to treat cable television as an independent medium with its own distinct set of regulatory problems. In 1972 the bureau was reorganized to include separate policy analysis divisions, and by 1974 the size of its staff had increased significantly. Through the creation and expansion of the new bureau, the FCC increased its analytical capabilities. The bureau became a centralized organizational mechanism for the collection, processing, and distribution of information and analysis concerning cable policy problems. Through the centralization of expertise, the bureau played a prominent role in the second economic inquiry and the repeal of major regulations.

FCC learning was also aided by a favorable institutional context. The Communications Act had given the FCC a broad grant of authority to regulate the communications industry, and although the authors of the Communication Act had not anticipated the advent of cable television, judicial deference allowed the FCC to implement an expansive reading of the act's statutory language. Confident by the late 1960s that it had the authority to regulate cable, the FCC felt free to experiment with different types of regulatory techniques. The APA rulemaking framework also helped create an institutional environment conducive to learning. By forcing the FCC to consider a broad range of evidence and to publicly justify its conclusions, the APA increased the likelihood that new information would be utilized when it became available. This institutional incentive structure by no means guaranteed learning, but it certainly made learning more likely.

Finally, the FCC was operating within a political context that was, at least in some respects, favorable to learning. On the one hand, the FCC confronted an increasingly complex and contentious interest group environment in the late 1960s and early 1970s. Fearing interest group resistance and lacking good information of its own, the FCC resorted to political bargaining when it formulated the 1972 cable rules. Yet even as the FCC faced pressure from numerous organized interests, it received little direct interference from Congress. Congress held oversight hearings that at times became contentious, but took no action to reverse FCC decisions. Indeed, the only cable television legislation Congress seriously considered was intended to enhance existing FCC powers. Ultimately, the technical complexity and political conflict surrounding cable television caused a legislative stalemate, placing responsibility for the problem's resolution squarely on the FCC's shoulders. These political conditions combined to allow the commission to exhibit increased decision-making autonomy as it gained knowledge and experience with cable television. Acting according to its own best judgment, the FCC revised and repealed regulations even when powerful interest groups voiced strong objections.

The findings in this study demonstrate that organizational learning was an important factor in the development of FCC cable television regulation in the 1970s. By 1980, the FCC had learned that most regulations designed to protect broadcasters from suffering economic harm due to cable television were unwarranted. That year, in fact, the FCC repealed the distant-signal carriage and syndicated exclusivity rules—two of the most important regulatory protections put in place in 1972. For this reason, 1980 stands out as a milestone in the history of cable television regulation and was purposely chosen as the endpoint of this study.

However, the deregulatory actions of 1980 did not mark an endpoint in the FCC's efforts to provide broadcasters with some regulatory protection from cable television, nor did they mark a complete end to the FCC's efforts to learn about the economic relationship between broadcasting and cable television. Furthermore, if the 1970s was in fact a time when substantial new knowledge was encoded within the FCC's organizational memory, the influence of this knowledge should be detectable in future policymaking endeavors.

For all of these reasons, it is useful to conclude this study with a brief discussion of the FCC's post-1980 regulatory activities.

Cable and Broadcasting: Regulation After 1980

During the 1980s, the cable television industry experienced a period of rapid growth, evolving into a major player in the communications marketplace. By the end of the decade, more than eight thousand cable systems were in existence, serving in excess of forty-five million subscribers. Furthermore, cable systems were evolving from "mom and pop" enterprises into large national corporations. By 1988, the largest cable provider, Tele-Communications, Inc., served just over a quarter of the cable subscribers in the United States. The ten largest cable providers collectively served more than 60 percent of all cable subscribers. The once-fledgling industry was now an $11.4 billion business.[2]

The rapid growth and consolidation of the cable industry was fueled by two factors. First, by the early 1980s, cable programming was progressing beyond mere carriage of traditional broadcast signals; new programming options such as specialized networks and premium subscription channels were making cable more attractive to potential subscribers in urban areas.[3] Second, further deregulation in the mid-1980s made it possible for cable to develop in a more openly competitive environment. In 1984 Congress passed the Cable Communications Policy Act, which limited the ability of state and local governments to regulate rates or attach unreasonable conditions to cable franchises.[4] With a uniform national regulatory structure in place, cable operators were able to invest and grow their enterprises in a more predictable (and favorable) environment.

With the rapid growth and consolidation of cable's economic power, federal regulators reconsidered the need for rules to protect and support broadcasters. In particular, the need for the must-carry and syndicated exclusivity rules was reexamined by the FCC in concert with the courts and Congress.

The Fall and Rise of Must-Carry

Following the completion of the economic inquiry in 1979, the FCC had moved to repeal the distant-signal carriage rules but had left the must-carry requirements in place. The decision to continue must-carry was controversial among cable programmers and small operators, who viewed the rule as an infringement of their First Amendment free speech rights. By guaranteeing broadcasters channel space on cable systems, cable interests argued, the FCC was essentially favoring one group of speakers over another and coercing speech on the part of cable operators. In 1980 Turner Broadcasting and Quincy Cable both petitioned the FCC to reconsider the must-carry requirements. Following a lengthy delay, the FCC denied the request, triggering a court challenge. In 1985 the US Court of Appeals sided with the cable industry, ruling that the must-carry rules were unconstitutional in their current form.[5]

The court concluded that the FCC could not infringe on cable operators' free speech rights without first articulating and defending an important governmental objective. However, in the 1979 economic inquiry, the FCC had presented substantial evidence demonstrating that regulatory protection of broadcasters was unwarranted, undermining the original rationale for must-carry in the process. Because the agency had continued to enforce the rules without clearly articulating any alternative rationale, it had inadvertently left the door open to the constitutional challenge. Although the court did not rule out the possibility that a revised version of must-carry might be constitutional, it made clear that the FCC needed to articulate and empirically defend a valid governmental goal for the policy.[6]

Facing pressure from both Congress and broadcasters, the FCC quickly opened a proceeding to consider the reinstatement of must-carry. Following an inquiry that lasted barely a year, the FCC reinstated must-carry rules on an interim basis—along with a new rationale for their existence: the maximization of consumer choice. The FCC claimed that in the long run, consumer choice in video programming was maximized by open competition—a situation that would not truly exist until broadcasters and cable providers could offer their services independently of one another. After twenty years of must-carry, however, video programming markets were distorted because many consumers relied on their ability to obtain broadcast signals via cable carriage. In order to level the playing field, the FCC proposed to reinstate must-carry for a five-year transitional period, during which cable operators would distribute channel-switching devices to their customers.[7] Dissatisfied with the commission's new rationale, cable operators once again took the FCC to court in 1987 and succeeded in getting the rules overturned.[8]

For the next five years, the cable industry remained free from mandatory signal carriage requirements. In 1992, however, must-carry requirements were reinstated by Congress as a part of the Cable Act of that year. This new iteration of must-carry required cable systems with twelve or more channels to devote one-third of their capacity to carriage of local broadcast signals. Local stations, for their part, were given the option of requesting mandatory carriage from the cable system or opting to negotiate retransmission consent in exchange for compensation from the cable system.[9] Seeking to inoculate the new rules against another court challenge, Congress carefully researched the issue and stated a number of justifications for must-carry in the text of the statute. Prominent among the stated rationales were the goals of (1) making local service widely available, (2) guaranteeing the stability of local-service origination, (3) promoting the continued availability of free broadcasting, and (4) ensuring an informed electorate.[10] Overall, Congress' stated interests in reinstituting must-carry seemed to reflect longstanding concerns for the preservation of broadcasting as a public interest industry.

Although the new statutory must-carry requirements were once again subjected to a constitutional challenge, this time they were upheld by the US Supreme Court.[11] Today, the 1992 must-carry provisions remain in effect.

The Return of Syndicated Exclusivity

In the late 1980s, the FCC also decided to reinstate an updated version of the syndicated exclusivity rules. After eight years of experience without the rules, the FCC concluded that the decision to repeal them had been "based on certain factual premises which with the benefit of hindsight, have proven to be … untrue."[12] In other words, the FCC in 1980 had failed to predict either the rapid expansion of cable or the explosion of cable programming that had subsequently occurred. When the rules were repealed in 1980, most cable program content had consisted of broadcast signal carriage. Thus, most syndicated content at the time was carried by broadcasters, who faced a modest economic threat if cable operators duplicated their programming by importing a distant signal. By the late 1980s, however, broadcasters of syndicated programming were in competition not only with network content but also with a variety of cable programming outlets. In this new economic environment, syndicated broadcast programming stood out as the only type of television content possessing no legal guarantee against non-duplication. The FCC believed that this put syndicated broadcast content at an economic disadvantage and opened an inquiry to consider reinstating syndicated exclusivity.

Mindful that the 1972 syndicated exclusivity rules were marred by excessive complexity, the FCC sought to implement "the simplest and most straightforward rules that will allow the marketplace for video programming to function evenhandedly."[13] The new rules treated all types of syndicated programming the same, extending a uniform period of protection that covered only the broadcasters' contract with program suppliers. In addition, the new syndicated exclusivity regulations applied uniformly to all television markets, regardless of size.[14] Confident that this updated formulation was administratively workable, the FCC ordered the rules into effect in 1988.

Concluding Thoughts

To sum up, organizational learning substantially influenced FCC cable television regulation during its formative years. By the end of the 1970s, FCC regulators had learned lessons about the broadcast-cable economic relationship and developed a new understanding of how cable fit into the broader structure of the US communications system. The events of the 1980s indicate that these lessons were in fact encoded in the FCC's organizational memory and that they continued to be influential throughout that decade. During that time, FCC policymaking reflected the belief that regulation was most desirable when it served to promote consumer welfare and protect the competitive marketplace. Accordingly, the reinstatement of the must-carry and syndicated exclusivity rules was justified as a way to enhance content diversity and promote effective competition between cable providers and broadcasters—justifications which reflect the second-order learning experienced by the commission some years earlier. The FCC also exhibited first-order learning when it reinstated the syndicated exclusivity rules; mindful of past failures, the commission drew on the lessons of experience in order to develop what it believed to be an improved version of the regulations.

This is not to say that learning was necessary in order for policy to change. Policy change might have occurred simply because cable interests became powerful enough to strong-arm Congress into overriding FCC regulations. Likewise, the courts might have favored a narrow reading of the Communications Act and struck down FCC regulations as overextensions of their statutory authority. But neither of these things happened. Indeed, it is precisely because they did not happen that learning occurred. Faced with a novel and highly complex policy issue, Congress and the courts preferred to defer to the FCC's judgment on numerous occasions. However, through oversight and enforcement of administrative procedures, these same institutions also demanded that the FCC "get cable right," so to speak. Interest groups, acting out of self-interest, continually put pressure on the commission to reconsider the status quo. These features of the FCC's immediate political and institutional context all made it possible for the agency to take action and experiment with policy solutions. With the right leadership in place, the commission took up the task of regulating cable television in the 1960s. In the 1970s, a combination of pragmatic leadership, experience, and increased organizational capacity allowed the FCC both to sharpen its understanding of the issues and to more decisively resolve longstanding policy controversies.

Thus, although learning was not necessary in order to produce policy change, it may very well have been necessary in order to produce more effective policy. Admittedly, it is difficult to say with any real precision which FCC cable regulations have been effective. For instance, policies such as must-carry and syndicated exclusivity persist in spite of a lack of evidence to demonstrate that they measurably advance the interests of broadcasters or consumers. Indeed, these policies both ceased to exist for several years in the 1980s, and broadcasters continued to survive and prosper. FCC learning in the 1970s, however, resulted in the permanent elimination of the leapfrogging, program origination, and distant-signal importation rules. Not only are these regulations gone, but the concerns that motivated their initial passage no longer constitute notable policy controversies. These changes suggest that regulators and interest groups alike have accepted a more limited role for regulation in the preservation of the national television broadcasting system. In this respect, at least, deregulation has proven to be a more effective policy strategy that the more restrictive rules of previous decades.

Do these findings suggest that regulatory agencies can and should become "learning organizations?" The answer is a qualified yes, at best. On the one hand, the findings of this study strongly suggest that organizational learning is a useful long-term strategy for improving government regulation. On the other hand, it seems unlikely that US politicians will ever purposely design a regulatory process that facilitates organizational learning. If politicians always delegated substantial decision-making autonomy to bureaucracies, hired pragmatic leaders and staff, and provided those individuals with the resources to gain expertise about policy problems, learning might become the norm in US regulatory policy. In reality, however, politicians rarely do all these things. The history of American regulatory policy demonstrates that politicians *do* sometimes design regulatory processes to favor certain outcomes. Furthermore, organizational leaders are more often chosen for party affiliation than for pragmatism, and organizational resources only endure as long as leaders and politicians find them beneficial. This does not mean that organizational learning is necessarily absent or even anomalous in

American regulatory policy. It does suggest, however, that politicians are unlikely to orchestrate the optimal conditions for organizational learning.

Indeed, with respect to FCC cable television regulation, organizational learning was by no means planned. Although Congress provided the FCC with a general and ultimately flexible statutory framework, it never anticipated that the agency would confront a problem as novel as cable television. Nor, for that matter, could Congress have predicted that the courts would uphold the broad interpretation of the Communications Act that made cable regulation possible. As organizational leaders, Dean Burch and Richard Wiley demonstrated unusual pragmatism at the FCC. However, like FCC chairmen before and after them, Burch and Wiley were chosen more for their party loyalty than for their personal characteristics or knowledge of communications policy. For this reason, their pragmatic leadership was probably not fully anticipated by the politicians who appointed them. Finally, organizational capacity came late to the FCC and was provided at the request of the FCC leadership rather than as a part of a congressional or presidential plan. Thus, as important as learning was to the formulation of FCC cable television regulation, it appears that politicians did not intentionally design the conditions that made cable television learning possible.

The findings of this study do, however, suggest that a regulatory agency can from time to time become a limited learning organization. That is, for brief periods of time, agencies may become learning organizations with respect to some issues within their jurisdictions. This is particularly true when agencies confront novel and technically complex policy problems that do not lend themselves to quick and easy solutions. Problems of this nature require agencies to seek out new knowledge and experiment with multiple policy alternatives over a period of years and decades.

When experimentation leads to repeated failures, growing public pressure may force agencies to either arrive at workable solutions or risk having solutions, workable or not, imposed upon them. Faced with these pressures, even politically appointed leaders will have incentives to improve organizational capacities and consider a broader range of policy instruments. Whereas such conditions do not guarantee the formulation of effective policy, they do provide the context within which issue-specific learning is likely to take place.

ENDNOTES

1. This case has been made with respect to a number of federal regulatory agencies. See Michael J. Zarkin, "Microeconomic Ideas, Policy Epistemologies, and the Politics of Spectrum Licensing, 1922–1997," *Polity* 38 (2006): 174–193. While other authors have not used the phrase "epistemic community," their conclusions support the thesis that a broader community of ideas was infiltrating regulatory practice at this time. See Martha Derthick and Paul J. Quirk, *The Politics of Deregulation* (Washington, DC: Brookings Institution, 1985); Marc Allen Eisner, *Antitrust and the Triumph of Economics: Institutions, Expertise, and Policy Change* (Chapel Hill: University of North Carolina Press, 1991).
2. Harry M. Shooshan III, "Cable Television: Promoting a Competitive Industry Structure," in *New Directions in Telecommunications Policy*, ed. Paula R. Newberg, 226–228 (Durham, NC: Duke University Press, 1989).
3. Ibid., 228.
4. For a thorough discussion of the law, see Robert F. Copple, "Cable Television and the Allocation of Regulatory Power: A Study in Governmental Demarcation and Roles," *Federal Communications Law Journal* 44 (1991): 1–172.
5. *Quincy Cable v. FCC*, 768 F.2d 1434 (1985).
6. Ibid., 1459.
7. *Mandatory Signal Carriage*, Report and Order, 1 FCC Rcd. 864 (1986).
8. *Century Communications v. FCC*, 835 F.2d 292 (1987).
9. The legislative history of the 1992 Cable Act does not specify the rationale for providing broadcasters with the option of both must-carry and retransmission consent. However, the issue here seems to be one of market power. The assumption is that cable systems will want to carry local stations with desirable programming (i.e., network affiliates). By declaring the intention to negotiate retransmission consent, these stations can leverage their market power to obtain monetary compensation from the cable operator. At the

same time, stations that declare the intent to negotiate retransmission consent are not counted toward the one-third of their channels that cable systems must make available for must-carry. In larger markets with many broadcast stations, this leaves more channel space available to smaller independent stations that might wish to declare must-carry privileges. On the issue of must-carry in the 1992 Cable Act, see David Swan, "'Must-Carry' in the 1992 Cable Bill: Will the Third Time be a Charm?," *Journal of Law and Commerce* 12 (1993): 367–393. On the issue of retransmission consent in the 1992 Cable Act, see Charles Lubinsky, "Reconsidering Retransmission Consent: An Examination of the Retransmission Consent Provisions of the 1992 Cable Act," *Federal Communications Law Journal* 49 (1996): 99–165.

10. *Cable Television Consumer Protection and Competition Act of 1992*, Public Law 385, 102nd Cong., 2nd Sess. (October 5, 1992), 2.
11. *Turner Broadcasting v. FCC*, 520 US 180 (1997).
12. *Program Exclusivity*, Report and Order, 3 FCC Rcd. 5299, 5303 (1988).
13. Ibid., 5313.
14. Ibid., 5314.

Bibliography

Secondary Sources

Aberbach, Joel D. *Keeping a Watchful Eye.* Washington, DC: Brookings Institution, 1990.

Argyris, Chris, Robert Putnam, and Diana McLain Smith. *Action Science.* San Francisco: Jossey-Bass, 1985.

Argyris, Chris, and Donald A. Schon. *Organizational Learning: A Theory of Action Perspective.* Reading, MA: Addison-Wesley, 1978.

Barnett, Harold J., and Edward Greenberg. "On the Economics of Wired City Television." *American Economics Review* 58 (1968): 503–508.

Bateson, Gregory. "Social Planning and the Concept of Deutero-Learning." In *Steps to an Ecology of Mind*, 159–176. New York: Ballantine, 1972.

———. *Steps to an Ecology of Mind.* New York: Ballantine Books, 1972.

Baugham, James L. *Television's Guardians: The FCC and the Politics of Programming, 1958–1967.* Knoxville: University of Tennessee Press, 1985.

Baumgartner, Frank R., and Bryan D. Jones. *Agendas and Instability in American Politics.* Chicago: University of Chicago Press, 1993.

Bawn, Kathleen. "Political Control versus Expertise: Congressional Choices about Administrative Procedures." *American Political Science Review* 89 (1995): 62–73.

Benjamin, Louise. "Working it out Together: Radio Policy from Hoover to the Radio Act of 1927." *Journal of Broadcasting and Electronic Media* 42 (1998): 221–236.

Bennett, Andrew. *Condemned to Repetition?: The Rise, Fall, and Reprise of Soviet-Russian Military Interventionism, 1973–1996.* Cambridge, MA: MIT Press, 1999.

Bensman, Marvin R. *The Beginning of Broadcast Regulation in the Twentieth Century.* Jefferson, NC: McFarland and Co., 2000.

Berner, Richard Olin. *Constraints on the Regulatory Process: A Case Study of Regulation of Cable Television.* Cambridge, MA: Harvard University Press, 1975.

Bernstein, Marver H. *Regulating Business by Independent Commission.* Princeton, NJ: Princeton University Press, 1955.

Besen, Stanley M., and Robert W. Crandall. "The Deregulation of Cable Television." *Law and Contemporary Problems* 44 (1981): 77–124.

Besen, Stanley M., Willard G. Manning, Jr., and Bridger M. Mitchell. "Copyright Liability for Cable Television: Compulsory Licensing and the Coase Theorem." *Journal of Law and Economics* 21 (1978): 67–95.

Brinson, Susan L. "Epilogue to the Quiz Show Scandal: A Case Study of the FCC and Corporate Favoritism." *Journal of Broadcasting and Electronic Media* 47 (2003): 276–288.

Broadcasting. "Cable Opts for Networking Role," 76, no. 26 (1969): 71–72.

———. "Chance for Accord on Cable's Future?," 76, no. 22 (1969): 23–26.

———. "Cox Appointment gives FCC a Club Flush," 63, no. 25 (1962): 58–59.

———. "FCC Starts Work on Basic Cable Policy," 75, no. 7 (1968): 58–59.

———. "Forget Your CATV Fears, Loevinger Says," 68, no. 24 (1965): 74–75.

———. "Heat in the Ashes of a Doused Deal," 76, no. 26 (1969): 68–69.

———. "In Sight: Stiffest Fight Yet Over Cable," 78, no. 21 (1970): 21–24, 26.

———. "Making Life a Bit Easier: Re-Regulation Gets Underway," 83, no. 19 (1972): 19.

———. "Mass Action Set for CATV," 71, no. 18 (1966): 52.

———. "Scratching the Itch to Originate," 76, no. 26 (1969): 69–70.

———. "Thumbs Down on the Cable Agreement," 76, no. 23 (1969): 24–26, 30.

———. "Two Blueprints for Unshackling CATV," 80, no. 17 (1971): 21–23.

———. "Where All the Talk about Cable May Lead," 80, no. 13 (1971): 81–82, 84–86.

———. "Wrecking Crews Hit Cable Compromise," 76, no. 24 (1969): 44–46.

Brown, M. Leann. "Learning and Food Security in the European Union." In *Organizational Learning in the Global Context*, edited by M. Leann Brown, Michael Kenney, and Michael Zarkin, 21–40. Aldershot, U.K.: Ashgate, 2006.

———. "Scientific Uncertainty and Learning in European Union Environmental Policymaking." *Policy Studies Journal* 28 (2000): 576–596.

Brown, M. Leann, and Michael Kenney. "Organizational Learning: Theoretical and Methodological Considerations." In *Organizational Learning in the Global Context*, edited by M. Leann Brown, Michael Kenney, and Michael Zarkin, 1–17. Aldershot, U.K.: Ashgate, 2006.

Brown, M. Leann, Michael Kenney, and Michael Zarkin, eds. *Organizational Learning in the Global Context.* Aldershot, U.K.: Ashgate, 2006.

Buck, Solon. *The Granger Movement: A Study of Agricultural Organization and its Political, Economic, and Social Manifestations, 1870–1880.* Cambridge, MA: Harvard University Press, 1913.

Carpenter, Daniel P. *The Forging of Bureaucratic Autonomy: Reputations, Networks, and Policy Innovation in Executive Agencies, 1862–1928.* Princeton, NJ: Princeton University Press, 2001.

Cohen, Wesley M., and Daniel A. Levinthal. "Absorptive Capacity: A New Perspective on Learning and Innovation." *Administrative Science Quarterly* 35 (1990): 128–152.

Copple, Robert F. "Cable Television and the Allocation of Regulatory Power: A Study in Governmental Demarcation and Roles." *Federal Communications Law Journal* 44 (1991): 1–172.

Daft, Richard L., and Karl E. Weick. "Toward a Model of Organizations as Interpretive Systems." *Academy of Management Review* 9 (1984): 284–295.

Derthick, Martha, and Paul J. Quirk. *The Politics of Deregulation.* Washington, DC: Brookings Institution, 1985.

Dodd, Lawrence C. "Learning to Learn: the Political Mastery of US Senators." *Legislative Studies Section Newsletter* (1992): 10–11.

———. "Political Learning and Political Change: Understanding Development Across Time." In *The Dynamics of American Politics*, edited by Lawrence C. Dodd and Calvin Jillson, 331–364. Boulder, CO: Westview Press, 1994.

Eisner, Marc Allen. *Antitrust and the Triumph of Economics: Institutions, Expertise, and Policy Change.* Chapel Hill: University of North Carolina Press, 1991.

———. *From Warfare State to Welfare State: World War I, Compensatory State Building, and the Limits of the Modern Order.* University Park: Pennsylvania State University Press, 2000.

———. *Regulatory Politics in Transition.* 2nd ed. Baltimore: Johns Hopkins University Press, 2000.

Ellis, Henry Carlton. *The Transfer of Learning.* New York: Macmillan, 1965.

Epstein, David, and Sharyn O'Halloran. *Delegating Powers.* Cambridge: Cambridge University Press, 1999.

Estes, William Kaye. *Learning Theory and Mental Development.* New York: Academic Press, 1970.

Etheredge, Lloyd S. "Government Learning: An Overview." In *The Handbook of Political Behavior.* vol. 2, edited by Samuel S. Long, 73–161. New York: Plenum Press, 1981.

Ferejohn, John A., and Charles R. Shipan. "Congress and Telecommunications Policymaking." In *New Directions in Telecommunications Policy*, edited by Paula Newburg, 301–314. Durham, NC: Duke University Press, 1989.

———. “Congressional Influence on Administrative Agencies: A Case Study of Telecommunications Policy.” In *Congress Reconsidered.* 4th ed., edited by Lawrence C. Dodd and Bruce I. Oppenheimer, 393–410. Washington, DC: Congressional Quarterly Press, 1989.

Fischer, Frank. *Reframing Public Policy: Discursive Politics and Deliberative Practices.* Oxford: Oxford University Press, 2003.

Flannery, Gerald V., ed. *Commissioners of the FCC: 1927–1994.* Lanham, MD: University Press of America, 1995.

Glaeser, Martin. *Outlines of Public Utility Regulation.* New York: Macmillan, 1927.

Glasbergen, Peter. “Learning to Manage the Environment.” In *Democracy and the Environment: Problems and Prospects*, edited by William Lafferty and James Meadowcroft, 175–193. Cheltenham, U.K.: Edward Elgar, 1996.

Gormley, William T. *The Politics of Public Utility Regulation.* Pittsburgh, PA: University of Pittsburgh Press, 1983.

———. “Regulatory Issue Networks in a Federal System.” *Polity* 18 (1986): 595–620.

Gottweis, Herbert. *Governing Molecules: The Discursive Politics of Genetic Engineering in Europe and the United States.* Cambridge, MA: MIT Press, 1998.

Haas, Peter. “Epistemic Communities and International Policy Coordination.” *International Organization* 46 (1992): 1–36.

Hajer, Maarten A. *The Politics of Environmental Discourse: Ecological Modernization and the Policy Process.* Oxford: Oxford University Press, 1995.

Hall, Peter A. "Policy Paradigms, Social Learning, and the State: The Case of Economic Policymaking in Britain." *Comparative Politics* 25 (1993): 275–296.

Harris, Richard A., and Sidney M. Milkis. *The Politics of Regulatory Change: A Tale of Two Agencies.* New York: Oxford University Press, 1989.

Hazlett, Thomas. "The Curious Evolution of Natural Monopoly Theory." In *Unnatural Monopolies: The Case for Deregulating Public Utilities*, edited by Robert W. Poole, Jr., 1–25. Lexington, MA: Lexington Books, 1985.

Heclo, Hugh. *Modern Social Politics in Britain and Sweden: From Relief to Income Maintenance.* New Haven, CT: Yale University Press, 1974.

Henck, Fred W., and Bernard Strassburg. *The Slippery Slope: The Long Road to the Breakup of AT&T.* Westport, CT: Greenwood Press, 1988.

Hooks, Gregory. "From an Autonomous to a Captured State Agency: The Decline of the New Deal in Agriculture." *American Sociological Review* 55 (1990): 29–43.

Houghton, David Patrick. "The Role of Analogical Reasoning in Novel Foreign Policy Situations." *British Journal of Political Science* 26 (1996): 523–552.

Huber, John D., and Charles R. Shipan. "Politics, Delegation, and Bureaucracy." In *The Oxford Handbook of Political Economy*, edited by Barry R. Weingast and Donald A. Wittman, 256–272. Oxford: Oxford University Press, 2006.

Jachtenfuchs, Marcus. *International Policy-Making as a Learning Process?: The European Union and the Greenhouse Effect.* Aldershot, U.K.: Ashgate, 1996.

Kahn, Frank J., ed. *Documents of American Broadcasting.* 4th ed. Englewood Cliffs, NJ: Prentice-Hall, 1984.

Kerwin, Cornelius M. *Rulemaking: How Government Agencies Write Law and Make Policy.* 3rd ed. Washington, DC: Congressional Quarterly Press, 2003.

Kingdon, John W. "Agendas, Ideas, and Policy Change." In *New Perspectives on American Politics*, edited by Lawrence C. Dodd and Calvin Jillson, 215–229. Washington, DC: Congressional Quarterly Press, 1994.

Klyza, Christopher McGrory. *Who Controls Public Lands?: Mining, Forestry, and Grazing Policies, 1870–1990.* Chapel Hill: University of North Carolina Press, 1996.

———. "A Window of Autonomy: State Autonomy and the Forest Service in the Early 1900s." *Polity* 25 (1992): 173–196.

Krasnow, Erwin G., Lawrence D. Longley, and Herbert A. Terry. *The Politics of Broadcast Regulation.* 3rd ed. New York: St. Martin's Press, 1982.

Krugman, Dean, M. "FCC Commissioner, Legal Assistant, and Staff Perceptions of Cable TV." *Journalism Quarterly* 56 (1979): 3–8.

Landy, Marc K., Marc J. Roberts, and Stephen R. Thomas. *The Environmental Protection Agency: Asking the Wrong Questions.* New York: Oxford University Press, 1990.

LeDuc, Don R. *Cable Television and the FCC: A Crisis in Media Control.* Philadelphia: Temple University Press, 1973.

Lee, Robert E. *In the Public Interest: The Life of Robert Emmet Lee from the FBI to the FCC.* Lanham, MD: University Press of America, 1996.

Levitt, Barbara, and James G. March. "Organizational Learning." *Annual Review of Sociology* 14 (1988): 319–340.

Levy, Jack S. "Learning and Foreign Policy: Sweeping a Conceptual Minefield." *International Organization* 48 (1994): 279–312.

Lichty, Lawrence, W. "The Impact of FRC and FCC Commissioners' Backgrounds on the Regulation of Broadcasting." *Journal of Broadcasting* 6 (1962): 97–110.

Litfin, Karen T. *Ozone Discourses: Science and Politics in Global Environmental Cooperation.* New York: Columbia University Press, 1994.

Loeb, Hamilton G. "The Communications Act Policy Toward Competition: A Failure to Communicate." *Duke Law Journal* no. 1 (1978): 1–56.

Lowi, Theodore J. "American Business, Public Policy, Case Studies, and Political Theory." *World Politics* 16 (1964): 687–691.

———. *The End of Liberalism.* New York: W. W. Norton, 1969.

———. *The End of Liberalism.* 2nd ed. New York: W. W. Norton, 1979.

Lubinsky, Charles. "Reconsidering Retransmission Consent: An Examination of the Retransmission Consent Provisions of the 1992 Cable Act." *Federal Communications Law Journal* 49 (1996): 99–165.

Lupia, Arthur, and Mathew D. McCubbins. "Learning from Oversight: Fire Alarms and Police Patrols Reconstructed." *Journal of Law, Economics, and Organization* 10 (1994): 96–125.

March, James G., Lee Sproull, and Michael Tamuz. "Learning from Samples of One or Fewer." *Organization Science* 2 (1991): 1–13.

McCraw, Thomas K. *Prophets of Regulation.* Cambridge, MA: Belknap Press, 1984.

McCubbins, Mathew D. "The Legislative Design of Regulatory Institutions." *American Journal of Political Science*, 29 (1985): 721–748.

McCubbins, Mathew D., Roger G. Noll, and Barry R. Weingast. "Administrative Procedures as Instruments of Political Control." *Journal of Law, Economics, and Organization* 3 (1987): 243–277.

McCubbins, Mathew D., and Thomas Schwartz. "Congressional Oversight Overlooked: Police Patrols versus Fire Alarms." *American Journal of Political Science* 28 (1984): 165–179.

Meier, Kenneth J. *The Political Economy of Regulation: The Case of Insurance.* Albany: State University of New York Press, 1988.

Miller, George. *Railroads and the Granger Laws.* Madison: University of Wisconsin Press, 1971.

Moe, Terry M. "Control and Feedback in Economic Regulation: The Case of the NLRB." *American Political Science Review* 79 (1985): 1094–1116.

———. "The Politics of Bureaucratic Structure." In *Can the Government Govern?*, edited by John E. Chubb and Paul E. Peterson, 267–329. Washington, DC: Brookings Institution, 1989.

Nisbett, Richard, and Lee Ross. *Human Inference: Strategies and Shortcomings of Social Judgment.* Englewood Cliffs, NJ: Prentice-Hall, 1980.

Northouse, Peter G. *Leadership: Theory and Practice*. 2nd ed. Thousand Oaks, CA: Sage, 2001.

Oliver, Michael J. *Whatever Happened to Monetarism?: Social Learning and Macroeconomic Policymaking in the United Kingdom since 1979*. Aldershot, U.K.: Ashgate, 1997.

Oliver, Michael J., and Hugh Pemberton. "Learning and Change in 20th-Century British Economic Policy." *Governance* 17 (2004): 415–441.

Paglin, Max D., ed. *A Legislative History of the Communications Act of 1934*. New York: Oxford University Press, 1989.

Park, Rolla Edward. *Audience Diversion Due to Cable Television: A Statistical Analysis*. Santa Monica, CA: RAND, 1979.

———. *Potential Impact of Cable Growth on Television Broadcasting*. Santa Monica, CA: RAND, 1970.

Parsons, Patrick R. "Defining Cable Television: Structuration and Public Policy." *Journal of Communication* 39, no. 2 (1989): 10–26.

Pool, Ithiel de Sola. *Technologies of Freedom*. Cambridge, MA: Belknap Press, 1983.

Rein, Martin, and Donald Schon. "Frame-Reflective Policy Discourse." In *Social Sciences and Modern States: National Experiences and Theoretical Crossroads*, edited by Peter Wagner, Carol Hirschon Weiss, Bjorn Wittrock, and Hellmut Wollman, 262–289. New York: Cambridge University Press, 1991.

Riker, William H. "The Heresthetics of Constitution-Making: The Presidency of 1787, with Comments on Determinism and Rational Choice." *American Political Science Review* 78 (1984): 1–16.

Robinson, Glen O. "Commetary on 'Administrative Arrangements and the Political Control of Agencies': Political Issues of Structure and Process." *Virginia Law Review* 75 (1989): 483–498.

Rose, Richard. *Lesson-Drawing in Public Policy: A Guide to Learning Across Time and Space.* Chatham, NJ: Chatham House, 1993.

Rosenberg, Herbert H. "Program Content—A Criterion of Public Interest in FCC Licensing." *Western Political Quarterly* 2 (1949): 375–401.

Rowland, Willard D., Jr. "The Meaning of 'Public Interest' in Communications Policy—Part II: Its Implementation in Early Broadcast Law and Regulation." *Communications Law and Policy* 2 (1997): 363–396.

Sabatier, Paul A. "Policy Change Over a Decade or More." In *Policy Change and Learning: An Advocacy Coalition Approach*, edited by Paul A. Sabatier and Hank C. Jenkins-Smith, 13–39. Boulder, CO: Westview Press, 1993.

Sabatier, Paul A., and Hank C. Jenkins-Smith, eds. *Policy Change and Learning: An Advocacy Coalition Approach.* Boulder, CO: Westview Press, 1993.

Schildhause, Sol. "CATV—From the Regulatory Point of View." *Public Utilities Fortnightly* 25 (September 1969): 78–81.

Schneider, Anne L., and Helen M. Ingram. *Policy Design for Democracy.* Lawrence: University of Press of Kansas, 1997.

Schon, Donald. "Generative Metaphor: A Perspective on Problem-Setting in Social Policy." In *Metaphor and Thought.* 2nd ed., edited by Andrew Ortony, 137–163. Cambridge: Cambridge University Press, 1993.

———. *The Reflective Practitioner: How Professionals Think in Action*. New York: Basic Books, 1983.

Seiden, Martin H. *Cable Television USA: An Analysis of Government Policy.* New York: Praeger, 1972.

Shipan, Charles R. *Designing Judicial Review: Interest Groups, Congress, and Communications Policy.* Ann Arbor: University of Michigan Press, 1997.

Shooshan, Harry M. "Cable Television: Promoting a Competitive Industry Structure." In *New Directions in Telecommunications Policy*, edited by Paula R. Newberg, 223–246. Durham, NC: Duke University Press, 1989.

Sikkink, Kathryn. *Ideas and Institutions: Developmentalism in Brazil and Argentina.* Ithaca, NY: Cornell University Press, 1991.

Skocpol, Theda, and Kenneth Finegold. "State Capacity and Economic Intervention in the Early New Deal." *Political Science Quarterly* 97 (1982): 255–278.

Skowronek, Stephen. *Building a New American State: The Expansion of National Administrative Capacities, 1877–1920.* Cambridge: Cambridge University Press, 1982.

Slotten, Hugh R. *Radio and Television Regulation: Broadcast Technology in the United States, 1920–1960.* Baltimore: Johns Hopkins University Press, 2000.

Spence, David B. "Agency Policymaking and Political Control: Modeling Away the Delegation Problem." *Journal of Public Administration Research and Theory* 7 (1997): 199–219.

Steinbrunner, John D. *The Cybernetic Theory of Decision.* Princeton, NJ: Princeton University Press, 1974.

Stigler, George J. "The Theory of Economic Regulation." *Bell Journal of Economics and Management Science* 2 (1971): 3–31.

Stone, Alan. *Public Service Liberalism: Telecommunications and Transitions in Public Policy*. Princeton, NJ: Princeton University Press, 1991.

———. *Wrong Number: The Breakup of AT&T*. New York: Basic Books, 1989.

Streeter, Thomas. "The Cable Fable Revisited: Discourse, Policy, and the Making of Cable Television." *Critical Studies in Mass Communication* 4 (1987): 174–200.

———. *Selling the Air: Commercial Broadcasting in the United States*. Chicago: University of Chicago Press, 1996.

Swan, David. "'Must-Carry' in the 1992 Cable Bill: Will the Third Time be a Charm?" *Journal of Law and Commerce* 12 (1993): 367–393

Teske, Paul. *After Divestiture: The Political Economy of Regulation*. Albany: State University of New York Press, 1990.

Weingast, Barry R., and Mark Moran. "Bureaucratic Discretion or Congressional Control?" Regulatory Policymaking by the FTC." *Journal of Political Economy* 91 (1983): 765–800.

Williams, Bruce A., and Albert R. Matheny. *Democracy, Dialogue, and Environmental Disputes: The Contested Languages of Social Regulation*. New Haven, CT: Yale University Press, 1995.

Williams, Wenmouth. "Impact of Commissioner Background on FCC Decisions, 1962–1975." *Journal of Broadcasting* 20 (1976): 239–260.

Wilson, James Q. *Political Organizations.* New York: Basic Books, 1973.

Wood, B. Dan, and Richard W. Waterman. "The Dynamics of Political Control of the Bureaucracy." *American Political Science Review* 85 (1991): 801–828.

Wood, B. Dan, and John Bohte. "Political Transaction Costs and the Politics of Administrative Design." *Journal of Politics* 66 (2004): 176–202.

Zarkin, Kimberly A., and Michael J. Zarkin. *The Federal Communications Commission: Front Line in the Culture and Regulation Wars.* Westport, CT: Greenwood Press, 2006.

Zarkin, Michael J. "The Evolution of Bureaucratic Design: The Development of the Federal Communications Regulatory Bureaucracy, 1926–1952." Paper presented at the annual meeting of the Western Political Science Association, Albuquerque, NM, March 2006.

———. "Microeconomic Ideas, Policy Epistemologies, and the Politics of Spectrum Licensing, 1922–1997." *Polity* 38 (2006): 174–193.

———. "Organizational Learning in Novel Policy Situations: Two Cases of United States Communications Regulation." *Policy Studies* 29 (2008): 87–100.

———. *Social Learning and the History of US Telecommunications Policy, 1900–1996: Creating the Telecommunications Act of 1996.* Lewiston, NY: Edwin Mellen Press, 2003.

———. "Telecommunications Policy Learning: The Case of the FCC's Computer Inquiries." *Telecommunications Policy* 27 (2003): 283–299.

PUBLIC DOCUMENTS

President's Task Force on Communications Policy. *Final Report.* Washington, DC: GPO, 1968.

Seiden, Martin H. *An Economic Analysis of Community Antenna Television Systems and the Television Broadcasting Industry.* Washington, DC: GPO, 1965.

US Congress. House. Committee on Government Operations. *Reorganization Plan No. 1 of 1970: Hearings before a Subcommittee.* 91st Cong., 1st Sess., 1970

———. Committee on Interstate and Foreign Commerce. *All Channel Television Receivers.* 87th Cong., 2nd Sess., 1962. H. Rep. 1559.

———. Committee on Interstate and Foreign Commerce. *Regulation of Community Antenna Television: Hearings before the Subcommittee on Communications and Power.* 89th Cong., 1st Sess., 1965.

———. Committee on Interstate and Foreign Commerce. Subcommittee on Communications. *Cable Television: Promise versus Regulatory Performance.* 94th Cong., 2nd Sess., 1976, Committee Print.

———. Committee on Interstate and Foreign Commerce. *Cable Television Regulation Oversight- Part II: Hearings before the Subcommittee on Communications.* 94th Cong., 2nd Sess., 1976.

US Congress. Senate. *A Message from the President of the United States Recommending that Congress Create a New Agency to be Known as the Federal Communications Commission.* 73rd Cong., 2nd Sess., 1934, S. Doc. 144.

———. Committee on Commerce. *Nominations—1969: Hearings before the Committee on Commerce.* 91st Cong., 1st Sess., 1969.

———. Committee on Commerce. *Appointments to the Regulatory Agencies: The Federal Communications Commission and the Federal Trade Commission.* Report prepared by James M. Graham and Victor H. Kramer. 94th Cong., 1st Sess., 1976. Committee Print.

———. Committee on Commerce. *Nominations June–October: Hearings before the Committee on Commerce.* 93rd Cong., 2nd Sess., 1974.

———. Committee on Commerce. *Nominations July–September: Hearings before the Committee on Commerce.* 94th Cong., 2nd Sess., 1976.

———. Committee on Interstate and Foreign Commerce. *Television Inquiry—Part VI: Hearings before the Committee on Interstate and Foreign Commerce.* 85th Cong., 2nd Sess., 1958.

———. Committee on Interstate and Foreign Commerce. *The Problem of Television Service for Smaller Communities.* Report prepared by Kenneth A. Cox. 85th Cong., 2nd Sess., 1958. Committee Print.

———. Committee on Interstate and Foreign Commerce. *Nomination of Newton Minow to be a Member of the Federal Communications Commission: Hearings before the Committee on Interstate and Foreign Commerce.* 87th Cong., 1st Sess., 1961.

US Federal Communications Commission. *Eighteenth Annual Report to Congress.* Washington, DC: GPO, 1952.

———. *Twenty-Second Annual Report to Congress.* Washington, DC: GPO, 1956.

———. *Twenty-Fifth Annual Report to Congress.* Washington, DC: GPO, 1959.

———. *Thirty-Ninth Annual Report to Congress.* Washington, DC: GPO, 1973.

———. *Fortieth Annual Report to Congress.* Washington, DC: GPO, 1974.

———. *Forty-Fifth Annual Report to Congress.* Washington, DC: GPO, 1979.

US Federal Radio Commission. *Third Annual Report to Congress.* Washington, DC: GPO, 1929.

Legal Citations

US Congress. *Cable Television Consumer Protection and Competition Act of 1992*. Public Law 385. 102nd Cong., 2nd Sess. (October 5, 1992).

US Federal Communications Commission. *Cable Television Report and Order*. 36 FCC 2d 141 (1972).

———. *Carter Mountain*. 32 FCC 459 (1962).

———. *Commission on Cable Television of the State of New York*. 43 FCC 2d 826 (1973).

———. *Deregulation of Radio*. 73 FCC 2d 457 (1979).

———. *Economic Impacts Inquiry*. Report. 26 FCC 403 (1959).

———. *En Banc Programming Inquiry*. 44 FCC 2303, 2312 (1960).

———. *First Illinois Cable TV*. 45 FCC 2d 304 (1974).

———. *First Report and Order in Docket 19995*. 52 FCC 2d 519 (1975).

———. *Frontier Broadcasting*. 24 FCC 251 (1958).

———. *In the Matter of Editorializing by Broadcast Licensees*. 13 FCC 1246 (1949).

———. *In the Matter of Mayflower Broadcasting Corp*. 8 FCC 339, 341 (1941).

———. *Intermountain Microwave*. 24 FCC 54 (1958).

———. *J. E. Belknap*. 18 FCC 642 (1954).

———. *Madison County Cablevision*. 42 FCC 2d 969 (1973).

———. *Mandatory Signal Carriage*. Report and Order. 1 FCC Rcd. 864 (1986).

———. *Memorandum Opinion and Order in Docket 18397*. 22 FCC 2d 825 (1970).

———. *Midwest Television* 13 FCC 2d 478 (1968).

———. *Notice of Inquiry in Docket 21284*. 65 FCC 2d 9 (1977).

———. *Notice of Inquiry and Notice of Proposed Rulemaking in Docket 15971*. 30 FR 6078 (1965).

———. *Notice of Inquiry and Notice of Proposed Rulemaking in Docket 19995*. 46 FCC 2d 1164 (1974).

———. *Notice of Proposed Rulemaking in Docket 18397*. 15 FCC 2d 417 (1968).

———. *Notice of Proposed Rulemaking in Docket 20487*. 53 FCC 2d 175 (1975).

———. *Notice of Proposed Rulemaking and Notice of Inquiry in Docket 19988*. 46 FCC 2d 139 (1974).

———. *Program Exclusivity*. Report and Order. 3 FCC Rcd. (1988).

———. *Report in Docket 20988*. 71 FCC 2d 951 (1979).

———. *Report in Docket 21284*. 71 FCC 2d 632 (1979).

———. *Report and Order in Docket 19988*. 49 FCC 2d 1040 (1974).

———. *Report and Order in Docket 20487*. 57 FCC 2d 625 (1976).

———. *Report and Order in Dockets 14895 and 15233*. 38 FCC 683 (1965).

———. *Report and Order in Dockets 20988 and 21284*. 79 FCC 2d 663 (1980).

———. *Second Further Notice of Proposed Rulemaking in Docket 18397-A*. 24 FCC 2d 580 (1970).

———. *Second Report and Order*. 2 FCC 2d 725 (1966).

———. *Sixth Report and Order*. 41 FCC 148, 167 (1952).

———. *Subscription TV Report and Order*. 15 FCC 2d 466 (1968).

US Circuit Court of Appeals. *Century Communications v FCC*. 835 F.2d 292 (1987).

———. *Home Box Office v. FCC*. 567 F.2d 9 (1977).

———. *Quincy Cable v. FCC*. 768 F.2d 1434 (1985).

US Supreme Court. *FCC v. Midwest Video Corp*. 440 US 689 (1979).

———. *FCC v. Pottsville Broadcasting Co*. 309 US 134, 137 (1940).

———. *Fortnightly Corp. v. United Artists Television*. 392 US 390 (1968).

———. *Munn v. Illinois* 94 US 113 (1877).

———. *Turner Broadcasting v. FCC*. 520 US 180 (1997).

———. *US v. Midwest Video Corp*. 406 US 649 (1972).

———. *US v. Southwestern Cable Co*. 392 US 157 (1968).

Archival Materials

American Broadcasting Company. Comments. 24 May 1974. Docket 19988. FCC Docketed Case Files, RG 173; Washington, DC: National Archives.

———. Petition for Rulemaking. 16 October 1964. Docket 15971. FCC Docketed Case Files, RG 173; Washington, DC: National Archives.

Association of Maximum Service Telecasters. Comments. 3 April 1969. Docket 18397. FCC Docketed Case Files, RG 173; Washington, DC: National Archives.

———. Comments. 7 December 1970. Docket 18397-A. FCC Docketed Case Files, RG 173; Washington, DC: National Archives.

———. Comments. Docket 19995. FCC Docketed Case Files, RG 173; Washington, DC: National Archives.

Charles River Associates, Inc. *The Audience Revenue Relationship for Local Television Stations*. 4 December 1978. Docket 21284. FCC Docketed Case Files, RG 173; Washington, DC: National Archives.

Cox Cable Communications, Inc. Comments. 22 May 1974. Docket 19988. FCC Docketed Case Files, RG 173; Washington, DC: National Archives.

Feldman, N. E. *Cable Television: Opportunities and Problems in Local Program Origination*. 21 October 1970. Docket 18397-A. FCC Docketed Case Files, RG 173; Washington, DC: National Archives.

Fischer, Franklin M. *The Impact of CATV Competition on Local Television Service.* 26 October 1964. Dockets 14895 and 15233. FCC Docketed Case Files, RG 173; Washington, DC: National Archives.

Fischman, Leonard L., and Associates. *Evaluation of FCC August 5, 1971 Distant-Signal Proposals for Cable Television in Terms of Their Impact on Over-the-Air Broadcasting.* 14 October 1971. Docket 18397-A. FCC Docketed Case Files, RG 173; Washington, DC: National Archives.

Geller, Henry. Oral History. Denver, CO: The Cable Center.

Kinley, David. Oral History. Denver, CO: The Cable Center.

KWRB-TV. Protest and Petition for Reconsideration. 27 May 1959. Docket 12391. FCC Docketed Case Files, RG 173; Washington, DC: National Archives.

Metromedia, Inc. Comments. 24 May 1974. Docket 19988. FCC Docketed Case Files, RG 173; Washington, DC: National Archives.

Midwest Video. Comments. 22 October 1970. Docket 18397-A. FCC Docketed Case Files, RG 173; Washington, DC: National Archives.

Missoula TV Cable Co., Inc. Comments. 26 June 1958. Docket 12443. FCC Docketed Case Files, RG 173; Washington, DC: National Archives.

National Association of Broadcasters. Comments. 7 July 1958. Docket 12443. FCC Docketed Case Files, RG 173; Washington, DC: National Archives.

———. Comments. 7 December 1970. Docket 18397-A. FCC Docketed Case Files, RG 173; Washington, DC: National Archives.

———. Comments. 29 July 1975. Docket 20487. FCC Docketed Case Files, RG 173; Washington, DC: National Archives.

———. Comments Relative to Paragraphs 11–20 and 23–25 of Part III of the Notice of Proposed Rulemaking and Notice of Inquiry. 3 June 1969. Docket 18397. FCC Docketed Case Files, RG 173; Washington, DC: National Archives.

———. Reply Comments. 26 October 1964. Dockets 14895 and 15233. FCC Docketed Case Files, RG 173; Washington, DC: National Archives.

National Cable Television Association. Comments. 7 December 1970. Docket 18397-A. FCC Docketed Case Files, RG 173; Washington, DC: National Archives.

———. Comments. 24 May 1974. Docket 19988. FCC Docketed Case Files, RG 173; Washington, DC: National Archives.

———. Comments. Docket 19995. FCC Docketed Case Files, RG 173; Washington, DC: National Archives.

———. Comments. 30 July 1975. Docket 20487. FCC Docketed Case Files, RG 173; Washington, DC: National Archives.

———. Comments on Paragraphs 11-20 and 23-25 of Part III of the Notice of Proposed Rulemaking and Notice of Inquiry. 3 April 1969. Docket 18397. FCC Docketed Files, RG 173; Washington, DC: National Archives.

National Community Television Association. Additional Reply Comments in Opposition to Proposed Rulemaking. 14 December 1964. Dockets 14895 and 15233. FCC Docketed Case Files, RG 173; Washington, DC: National Archives.

———. Reply Comments. 28 July 1958. Docket 12443. FCC Docketed Case Files, RG 173; Washington, DC: National Archives.

Schildhause, Sol. Oral History. Denver, CO: The Cable Center.

Southern Idaho Broadcasting. Comments. 7 July 1958. Docket 12443. FCC Docketed Case Files, RG 173; Washington, DC: National Archives.

US Federal Communications Commission. Cable Television Bureau. *Impact of Cable Television on Audience Shares*. Docket 21284. FCC Docketed Case Files, RG 173; Washington, DC: National Archives.

———. CATV Task Force Chief. Inter-office Memorandum. 30 July 1968. Papers of Commissioner Rosel Hyde, MSS 142; Provo, UT: Brigham Young University Special Collections.

———. General Counsel and Chief, Broadcast Bureau. Inter-office Memorandum. 30 July 1968. Papers of Commissioner Rosel Hyde, MSS 145; Provo, UT: Brigham Young University Special Collections.

———. Memorandum Opinion and Order in Dockets 14895, 15233, and 15971. 20 January 1967. FCC Docketed Case Files, RG 173; Washington, DC: National Archives.

———. Notice of Inquiry and Notice of Proposed Rulemaking. 22 April 1965. Docket 15971. FCC Docketed Case Files, RG 173; Washington, DC: National Archives.

———. Notice of Proposed Rulemaking. 14 December 1962. Docket 14895. FCC Docketed Case Files, RG 173; Washington, DC: National Archives.

———. Panel Discussion. 11 March 1971. Docket 18397-A. FCC Docketed Case Files, RG 173; Washington, DC: National Archives.

———. Public Notice: Cable Television Bureau Releases Research Studies Relevant to Cable Inquiry. Docket 21284. FCC Docketed Case Files, RG 173; Washington, DC: National Archives.

———. Report and Order. 25 June 1956. Docket 11532. FCC Docketed Case Files, RG 173; Washington, DC: National Archives.

———. Statement of the Broadcast Bureau. Oral Argument. 14 December 1961. Docket 12391. FCC Docketed Case Files, RG 173; Washington, DC: National Archives.

Viacom International, Inc. Comments. 22 May 1974. Docket 19988. FCC Docketed Case Files, RG 173; Washington, DC: National Archives.

Wharton EFA, Inc. *Impact of Cable TV on Local Station Audience.* March 1978. Docket 21284. FCC Docketed Case Files, RG 173; Washington, DC: National Archives.

Index

Made in the USA
Monee, IL
12 November 2026

19d62f37-e6e8-4b75-b579-c09ec81e97f6R01